THE NATURE OF THINGS

"WILL CURTIS,

and this is
THE NATURE OF THINGS"

With an introduction by Les Line,
editor of Audubon Magazine

ILLUSTRATIONS BY NORA S. UNWIN

The Countryman Press, Woodstock, Vermont 05091

Copyright © 1984 by Will Curtis

All rights reserved. No part of this book may be reproduced in any form or by any electronic or mechanical means including information storage and retrieval systems without permission in writing from the publisher, except by a reviewer who may quote brief passages in a review.

Permissions and copyright notices for all contributed material are found on pages 283–291.

Engravings by Nora S. Unwin:

Pages 32, 58, 102, 141, 142, 168, 217: from *Footnotes on Nature* by John Kieran. Copyright 1947 by John Kieran. Reproduced by permission of Doubleday & Company.

Facing page 1; and pages 160, 179, 201, 238, 256, 272. Reproduced by permission of the Sharon Arts Center, Sharon, New Hampshire, and the Estate of Nora S. Unwin.

Library of Congress Cataloging in Publication Data
Main entry under title:

The nature of things.

   Includes index
   1. Natural history—Addresses, essays, lectures.
I. Curtis, Will, 1917–
QH81.N295   1984   508   84-17656
ISBN 0-88150-028-3

Printed in the United States of America.

To Jane, partner in acting, dairy farming, bookselling, writing, and radio broadcasting; and companion in traveling.

Contents

PREFACE ix
INTRODUCTION xi
THE SEASONS 1

Seasons • A Hillside in Spring • Spring Trees • Rite of Spring • Spring Sounds • Shadbush • Lilac Time • Leaving Wildlife Youngsters Wild • Dandelions • Dew • Hiking • Stars on a Summer Night • The Scythe • Bright Passage • The Maple Casebearer • Ponds Change in Autumn • Banking the House • Winter Bird Feeding • Snow Is Nature's Security Blanket • Snowshoes • Animals Keeping Warm • Winter Feet • Snow Fleas • A Long Winter's Nap • "Habitat" Gardening

PLANTS & GARDENS 33

How Soils Are Made • Roadside Flowers • Weeds: Foreign Friends or Alien Enemies? • Poisonous Plants • Creeping Menace • Ancestral Diet • Garibaldi's Garden • High-yield Vegetables • How to Make Better Compost • Manure Means Money • Witch Hazel • Garden Wrap-up • Home Storage of Vegetables • Leaf Drop of Evergreens • Autumn Leaves • Fall Lawn Care • Watering and Indoor Plants • Moss and Berry Bowls

MAMMALS 59

Animals • Growing Old • Yawning Animals • Animal Language • Insulation • Migration • The Hunter and the Hunted • Bag Balm • The Beaver • Red Squirrel • Antlers • The Red Deer • Himalayan Animals • The Buffalo Graze, Again • Mules • Meet Wildlife Enemy No. 2 • Coyotes • The Big, Good Wolf • A Flash of Mink • The Weasel • The Fisher • The Water Shrew • The Jaguarundi • The Cougar's New Cloak • Nature's Golden Racer: The Cheetah • The Wiley, Indigestible Armadillo • The Sea Otter • Animals' Diving Gear • The Tragic Facts About Whales • The Bowhead Whale • White Whale • Questions for a Blue Whale

BIRDS 103

Rites of Spring • Growing Up in a Nest • Masters of Adaptation • Spring Bird Migration • Feathers • How Birds Change Their Clothes • Man-made Bird Houses • "Mobbing" • Independent Nestlings • The Upside-Down Bird • Whip-Poor-Wills • Bluebirds • Cukoos Aren't Crazy, Just Lazy • Swifts • The Pine Siskin • Grouse • Raising a Great Horned Owl

• Ostriches • The Bald Eagle • Condors • Turkey Vultures • Kingfishers • The Common Loon • The Bittern • Whooping Cranes • Gulls • Puffins • Killdeer • A Guide for Gannet Gazers

INSECTS   143

Insect Structure • Insects in a Stream • Insect Antifreeze • A Bounty of Beetles • Beetlemania • Honeybees • Bumblebee • Bees as Undertakers • Bee Bites • Lightning Bugs • Grasshoppers • Crickets • Butterflies of Spring • The Monarch • Pill Bug • Aphids • Crab Spiders • Black Widow Spider's Bite

WATER & AQUATIC LIFE   169

Water • Water Tables • Woodlots and Water • Trends in Neurobiology • Leaking Gas Tanks • Wetlands • Swamp Bubbles • Peat • Tidepools • Estuary • Salamanders • Shells and Man • Starfish • Sea Urchins • Blue Lobsters? • Pickled Wrinkles • The Living Fossil • Sand Dunes • The Science of Oceanography • Salmon • Return to the Connecticut • The Seahorse • Shrimp Farming

PLACES   201

Canyon de Chelly • Early Prairie Homes • Yellowstone Backcountry • Mount Assiniboine • The Columbian Ground Squirrel • Caribou • Adirondack Guide Boat • The Cardiff Giant • Wine from the Finger Lakes • St. Brendan • Forgotten Viking Boat • Beatrix Potter • Laplanders • The Lipizzan Horses • The Tower of Pisa • Sardinia • Cordoba • The Galapagos Islands • Peruvian Textiles • Inca Empire • Quechua Indians • Llamas • Peruvian Food • Peru • Mountain Trails • Dead Woman Pass • Trek to Machu Picchu • Cuzco

PLANETS & SPACE   239

A Short History of All History • The Winter Sky • Arabian Astronomy • Lightning • Space Trash • Meteors and Meteorites • Asteroids • Getting Ready for Halley • Marsquake • Survival in Space • Aurora Borealis • The Technicolor® Sky • Project Sentinel

ENERGY   257

Energy and Agriculture • Solar Greenhouse • Wood Energy • Drying Wood • Wind Power • Wind-powered Generating Systems • Kerosene • Efficient Lighting and Small Appliances • Coal • Oil-platform Fires • Arctic Mining

THE NUCLEAR SWORD   273

Reduction of the Ozone Layer • A Darkened World • A Full-scale Nuclear War? • What About the Children? • Twisted Words, Distorted Images • The Ultimate Folly

CONTRIBUTORS   283

INDEX   293

Preface

> The goal of life is living in agreement with nature.
> —Zeno, 335 BC

Peering into a glass bowl to watch mice endlessly chasing each other's tails was my earliest introduction to animals. I was told they were "dancing mice."

By the time I had entered school, tame, white rats with beautiful pink eyes had taken the place of the mice. Of course, I had to show them off at school. They didn't remain hidden in my desk very long, for as soon as I removed the inkwell they came streaming out, amid the shrieks of my schoolmates.

In the early 1930s, ducks and geese still filled the sky over Cape Cod. Ours was a hunting family, and by the age of fourteen, a boy was supposed to know how to handle a double-barreled shotgun. It was no great feat to bag your daily limit, sometimes in a matter of minutes.

But to me it was a cruel, tragic thing when I shot my first and last goose and carried its beautiful but lifeless body back to camp. It was located on the great beach at Nauset and close to the cottage of Henry Beston, where he wrote his classic *The Outermost House*. I support fully the rights of respectful hunters and the value of the federal duck stamp that has raised many millions for the purchase of wetlands. Now on my walks, observing in fields and woods, I carry no gun.

As a dairy farmer, I learned much about birds flying and nesting from the seat of a tractor while harrowing or mowing. More than once, I raised the cutter bar in the nick of time to avoid hitting a hidden fawn. And, looking back over the hay I had cut, I sometimes saw foxes springing into the air to pounce on the elusive field mice.

When farming days were over, owning a bookstore brought many books about nature to my attention, eventually to be reviewed on the air. Commercial radio broadcasts led to public radio commentaries when I was president of the Vermont Institute of Natural Science (VINS) in Woodstock, Vermont, an extraordinarily useful private institution which is now widely known for its educational programs. Most successful and unique is the statewide Environmental Learning for the Future (ELF) program in which community volunteers are trained by the VINS staff as naturalists/teachers who work with children in the schools. Among other contributions to the awareness and knowledge of the natural world around us, VINS sponsors programs and courses for adults, such as bird and fern walks and teacher workshops. VINS is also engaged in ornithological research and has completed a Breeding Bird Atlas Project in Vermont, operates a bird banding station, conducts the Common Loon Survey, and is licensed to rehabilitate injured birds. From its Raptor Center came the live owl pictured with me on the cover of this book.

The knowledgeable VINS staff has contributed many of the commentaries collected here. The first year of what Betty Smith titled "The Nature of Things" over Vermont Public Radio was sponsored by VINS. For the past three years, thanks to a major grant from the National Audubon Society, the program has been relayed to a growing number of National Public Radio stations across the country by the mysterious (to me) means of a satellite 22,300 miles in the sky.

—Will Curtis

Introduction

Memories. Of life during the Golden Age of Radio. Of the giant Philco console that dominated the modest living room of our home in a western Michigan farm town. Of the spindly aerial that waved over the clothesline pole in the backyard, intercepting marvelous sounds from clear-channel stations in distant cities and shortwave broadcasters in unimaginably faraway lands.

Family hours in those years—before, during, and immediately after World War II—were spent around the radio, listening to our favorite weekly dramas and variety programs. From the huge speaker of that Philco came spine-tingling mysteries, tear-jerking romances, the uncanned laughter of live comedy, the swinging jazz of big bands and small combos, the war-horses of symphony orchestras. And news of a world aflame.

The hours before suppertime were for kids. For make-believe heroes whose adventures sold breakfast cereals. For special offers of rings with secret compartments that cost a boxtop and a dime—and took an interminably long time to arrive in the mail.

Summer afternoons were for baseball, for Bert Wilson's accounts of Chicago Cubs games from beautiful, vine-clad Wrigley Field. But if the Cubs were on the road, it would be a masterful fraud, a re-creation, complete with roaring crowd noises at appropriate moments, of a Western Union ticker-tape report. And in your imagination, Bert was right there at the Polo Grounds or Ebbets Field, watching the exploits of Cubs stars like Phil Cavarretta, Andy Pafko, and Hank Sauer.

Imagination. That was the essence of radio, before television. Four actors and actresses, a sound-effects man, and your imagination could instantly create the dusty main street of Dodge City, Kansas, in the 1870s. There, every Sunday afternoon, Marshal Matt Dillon, in the deep voice of William Conrad, kept law and order with the help of his stiff-legged deputy Chester, the encouragement of his saloon-keeping lady friend Miss Kitty, and the bark of his Colt six-shooter. And it worked, really worked. Why else, after thirty-five years, would a grown man remember, with a chill, the night he was left alone in a dark house to listen to "The Inner Sanctum," whose creaking door opened on a frightening tale of rubber plantation workers in the South American jungle fleeing an unstoppable horde of flesh-devouring army ants?

Television, of course, requires little work of the imagination, and in darker moments I wonder if this will one day destroy man's ability to create. Newton Minow called television "a vast wasteland," an accusation I can't challenge, and commercial radio in television's wake has become a barren moonscape. Easy-listening music is prepackaged in distant studios and syndicated to local stations that may be totally automated, without even a single live deejay on the payroll. Earthshaking events are briefly mentioned in sixty-second newscasts. A day's most exciting broadcast moment is the traffic helicopter's report of a monumental expressway tie-up at rush hour.

But all is not lost for those of us who grew up with radio. Who twist the dials of our car and bedside and kitchen sets, or our high-fidelity receivers, in the hope of finding programming that is entertaining, informative, and challenging. Commercial radio may be a moonscape, but the sound of quality can still be heard on the airwaves. Its name is Public Radio, and its *raison d'être* is substance, not profit. Its believers know that radio is not a medium of the distant past, that it can do things that television can't—or won't. And so Public Radio regularly brings us such rarities as Garrison Keillor and his witty "Prairie Home Companion," orchestras great and near-great, live jazz, the news and public affairs report "All Things Considered," Shakespeare, readings of best-selling books. Plus, on several dozen stations from Maine to Alaska and south to Texas, Will Curtis and "The Nature of Things."

"Thank you for the reassurance that some things are still the same," one listener wrote to Vermonter Will Curtis, whose soft-spoken commentaries on nature—and man's relationship to nature—are wafted nationwide by satellite from Vermont Public Radio, free for the asking. And to the National Audubon Society, which through an annual grant supports the production of "The Nature of Things," came this letter from Fairbanks, Alaska: "Thanks ever so much for the commentary on the threat of nuclear war and its psychological effects on children. We need more of this! The program was clear and direct. Maybe we can inject some sanity into world politics.

Those two letters give you some idea of the scope and inpact of "The Nature of Things," a three-minute, 365-days-a-year mini-series that is four years old this fall. Most often, Will Curtis will enrich listeners' awareness of the world about them by exploring the intricacies of nature—the reason wolves howl, for instance. Or why fireflies flash, the way ice is formed, how birds stay warm in the depth of winter. Some features will be helpful, such as a discussion of the best kinds of firewood. Others are nostalgic, like the story of the old stone walls that once bordered farmers' fields throughout New England, or a program on the noises in old barns. Often there are appropriate sounds in the background—the voices of birds and frogs, the roar of a geyser in Yellowstone, the fall of a tree in the forest. But Will does not eschew the serious concerns of environmentalists. And no threat is more serious than nuclear winter, which could destroy all life on Earth.

Will Curtis is uniquely qualified to be nature's interpreter and spokesman. He has spent his life close to the land—wandering the forests and marshes of Cape Cod as a boy, working farms in Massachusetts and Vermont. As a member of the Vermont Legislature, he served on the House Conservation Committee and supported laws against billboards and throwaway bottles and cans. And he was president of the Vermont Institute of Natural Science. Will and his wife

Jane have traveled widely, not just to wild places in America but to the Galapagos Islands, the Peruvian Andes, Egypt, Scotland, and Finland. And they co-authored a book on America's first environmentalist, George Perkins Marsh.

Four years of daily broadcasts means nearly fifteen hundred editions of "The Nature of Things," which means that Will Curtis is a voracious reader. For topics and research, he draws on many sources—books, of course, but especially magazine articles, and *Audubon* is a deep well of ideas. We're pleased that several of the two hundred essays in this anthology from "The Nature of Things" originated in our pages. And we're proud that the National Audubon Society has been able to bring Will Curtis every day of the year to thousands of Public Radio fans from New England to the Far North and Deep South.

—Les Line
Editor, *Audubon* Magazine

Publisher's note: Les Line became editor of Audubon *in 1966 and transformed it into what Roger Tory Peterson has called "not only the most beautiful natural history magazine in the world, but the most beautiful magazine of any sort in the English language." He is a noted photographer and co-author of several books.*

THE NATURE OF THINGS

The Seasons

SEASONS

What do we think of when we use the word, *seasons*? The dictionary gives its root as the Latin verb *serere*, meaning *to sow*. From there it took the meaning of a *season to sow* and on to our seasons. Today we use the word primarily with the idea of the four seasons; winter, spring, summer, and fall. But think of all the other ways it's used! Here are a few of them. There's a season when a product is available, say, the strawberry season or the oyster season. Then there's the season when an activity is possible, the deer season, the baseball season, the skiing season. There's a period with reference to the total number of games played by a team, a 158-game season. We also refer to a time in our lives, "the season of my childhood," and carpenters refer to "seasoned" lumber.

But to get back to the theme of seasons and the year, you doubtless have heard of the terrible year in the early part of the nineteenth century when there weren't four seasons, only one. It's still referred to as "Eighteen-sixteen and froze to death." Records show that snow fell every month of the year. Spring never came, summer never came, and in July snow was still piled up with more coming down. Crops never got in the ground, animals died from lack of food; humans went hungry. In Vermont, where conditions were particularly severe, whole families pulled up roots and left for parts of the country where such unnatural weather didn't plague honest, hardworking folk.

Our seasons are controlled by the patterns of weather moving across the country from west to east. Usually the patterns move in a more or less predictable fashion, but in 1816 the sun's warmth and its action on the weather was obscured by an enormous number of particles in the air. A volcano in the Far East had erupted and spewed out so much matter that it affected the amount of warmth reaching New England. So summer never arrived. Snow droughts in some recent years have been due to the fact that the jet streams that usually loop down and pick up moisture from the Mexican Gulf instead sped straight across the northern part of the continent, bringing lots of cold, dry air.

There are two other ways we use the word *season* that are peculiar to New England: how about "mud season" and "sugar season"? And with sugar season and its companion, mud season, some folks think of getting away for a change. I've heard a doctor say that our bodies take a lot of punishment from the changing seasons and that it is a good idea to take a vacation four times a year just to rest up from all that strain. Yet I always thought that we who lived in northern climates benefited from the changing seasons; we're supposed to be smarter, more alert, and more active than those who just lie about all day waiting for the fruit to drop from the trees into their open mouths. Well, I for one love the changing seasons. And to end this discussion I'll quote one of the most famous passages on the subject from Ecclesiastes: "To everything there is a season and a time to every purpose under the heaven." [1]

A HILLSIDE IN SPRING

As the spring sun continues to rise higher in the sky each day, its rays steadily work to melt the winter's accumulation of snow. The depth of the snow continually decreases and the area covered becomes less and less.

The first bare ground usually appears on south-facing hillsides. Because the sun is relatively low in the southern sky, its rays shine at a shallow angle across level ground. A hill may intercept the rays at close to a right angle, in effect giving the sun much more melting power.

From a distance, the first bare areas appear to be covered simply by brown, matted grass and weeds, killed by the long, cold winter. If you look closely, however, you will see that the short, center blades in each grass clump are not brown and dead, but green and alive. Running and creeping over and through the vegetation are the first spiders, mites, and insects of the new year, brought forth from hibernation by the sun's warmth.

Larger animals also take advantage of the first-thawed ground of spring. Large, black-plumaged crows walk slowly, stiff-leggedly across the hillside, stopping frequently to snatch up a tasty morsel. Early returning robins depend on bare hillsides as their hunting ground also, for it is difficult to locate earthworms under snowdrifts and frozen ground.

The receding snow also reveals a maze of tunnels and trails. Moles tunnel just under the surface, leaving a ridge of moist dirt. A small mound marks the place where one has ventured deeper, leaving the displaced soil in a pile on the surface. Meadow mice, actually voles, also make a network of tunnels—but theirs are on the surface, through the vegetation. If you find a mouse-sized tunnel through the grass, paved with a layer of short, cut-up grass stems, you have discovered a meadow vole runway.

You do not have to climb or look closely at the ground to observe white-tailed deer grazing on the hill. Deer spend the winter browsing on twigs and buds in nearby woods.

By late winter, many, particularly fawns and does, may be suffering from a shortage of food. When the first bare patches are melted on hillsides, deer eagerly move out to graze on the remains of last year's vegetation and the first green sprouts of spring. Small groups of deer may spend the entire day on the hill, if they are not disturbed by people or roving dogs. When new growth appears in the woods and fields, the deer spread out to more protected areas to feed. The hillside continues to become greener, as new growth covers the old, and summer arrives shortly. [2]

SPRING TREES

Many people enjoy spring for its many colorful wildflowers. While the bloodroot, hepaticas, violets, trilliums, and spring beauties are providing welcome color on the ground, many trees are providing another show of blossoms above our heads.

Pussy willows are traditionally one of the first signs of spring. In fact, a few may even blossom during a midwinter thaw, if the buds' cold requirements have been met. The poplars are in the same family as the willows, and like them are dioecious, that is, they have male and female flowers on separate trees. Most poplars are also early bloomers; their catkins emerge from the protective covering of the bud as early as March in some species. The new leaves, which are more susceptible to freezing temperatures on cold spring nights, remain safely in the buds for a while longer.

Many of the earliest bloomers of the tree world grow in swamps and around ponds, probably because water is first melted and available for growth in wetlands. In addition to the willows, red maple and tamarack usually begin flowering in March. The branches of the red maple bear tightly packed clusters of flowers that show up from a distance. Close inspection reveals the beauty of the individual flowers; the bud scales that protected the buds over the winter now have the appearance of red-tinted petals, and the many stamens dangling from the blossom look like miniature wind chimes. The blossom of the tamarack appears along with the new needles, producing a bright combination of colors—yellow, pollen-covered male flowers, oblong female flowers with brilliant crimson bracts, and the soft green of the growing needles. [3]

RITE OF SPRING

On the proper nights in late March or early April, throughout New England the salamanders emerge en masse to move through the dripping woodlands in a sort of dark, inverted rite of spring. Slowly, their pale bellies barely clearing the damp leaf litter, they will crawl toward the ponds where they were born. Once on the move they are single-minded in their intent, oblivious to predators, passing cars, roads, culverts, and overeager collectors from biological supply houses.

Once they reach the breeding ponds, the males will begin to drop spermatophores, the vaselike bodies containing the sperm. Later the females will arrive, and in what one herpetologist describes as an entirely sensuous ritual, the males will twist and thrash about in the water as they nudge the females toward the floating spermatophores. In the space of a night it is over. The females will pick up the spermatophores in their cloacae, the males will disperse, and a few days later eggs will be deposited on a submerged twig. Two to three weeks later, the eggs hatch into tiny, gilled salamander larvae that resemble tadpoles. By the end of the summer, the larvae will develop into smaller versions of the adults and leave the drying ponds to seek out some dark crevice in which to spend their lives.

The night migration of the salamanders is typical of any number of momentous events in natural history that take place under the very nose of the civilized human community. And yet, in spite of the fact that it has been calculated that the biomass of salamanders in the Northeast surpasses that of birds, the breeding ritual and migration go unnoticed by all but a few dedicated naturalists.

A combination of acid precipitation and destruction of habitat by suburban housing developments and shopping centers, as well as heavy pressure from biological supply houses that use the salamanders for study purposes, is decimating the populations of salamanders of the genus *Ambystoma*. Recently the pet trade has joined this host of enemies, evidently for no other purpose than to profit from unwary customers — more often than not salamanders die in captivity, and even when they survive they do little but hide beneath the soil of aquariums. [4]

SPRING SOUNDS

"Kraks," sleighbells, ear-piercing trills. From April on, evenings are full of the sounds of courting amphibians. Wood frogs are the first to be heard with their croaking quacks, then spring peepers join in, chorusing in high-pitched peeps. A little later, after the wood frogs are silent, American toads sing with an incessant, crescendo of trills.

The courtship singing is of course prelude to mating and egg-laying. Should you be a daytime outdoor person, you might miss these deafening songs of spring pond creatures, but you may well notice speckled gelatinlike blobs in the water and wonder what they are. Each spring I try again to learn and remember which eggs belong to which amphibians.

The spotted salamander, shiny black with bright yellow dots, is one of the earliest egg-layers, with eggs massed in a blob of jelly, attached to submerged twigs or stems; one author describes them as looking like tapioca pudding. The red-spotted newt, very common and plentiful in local ponds, lays eggs one at a time on a stem or leaf of a water plant.

Wood frogs are the earliest frogs to lay eggs. The eggs could be confused with spotted salamander egg clumps, except the egg mass is greener and much larger, two to four inches. Two to three thousand eggs may be in a single clump. These also are fastened to twigs and grasses at the pond's edges. Spring peepers lay their eggs one at a time on submerged plants; as one female may lay hundreds of eggs, this can take weeks. Leopard and pickerel frogs lay their eggs in bulky gelatinous masses, which float suspended in the water, while green frog and bullfrog eggs spread out over the surface like a piece of beaded net. As many as 20,000 black and white eggs may be contained in a mass of bullfrog eggs.

American toad eggs need not be confused with other amphibian eggs if you think of T for "toad" and "tube." American toads lay long tubes of jelly, each with a single row of eggs in it. Right after eggs are laid, water swells the gelatinous mass, and the tube thickens to perhaps a quarter inch in diameter.

Identifying egg masses in ponds may not be your top priority, but it's fun to know what you're looking at! [5]

SHADBUSH

Every year we say this is one of the most beautiful springs we've ever had. After the long steel-cold winter and the endless days of mudtime, the arrival of spring with the birds goes right to our hearts. And what is the most beautiful sight of all is the ephemeral loveliness of the shadbush, also known as shadblow, or in the South as serviceberry. Shad is a shrub, that is, a woody plant less than twelve feet tall having more than one stem. Shad is widespread in the eastern United States; there are four or five species, and in northern New England some refer to it as our dogwood although of course it is quite different. What I think is meant is that it produces the same gasp of admiration as one sees the fragile white loveliness among the green of the poplars and maples on our hillsides. The flowers appear before the leaves have fully unfolded, a cluster of five narrow white petals. Later purplishblack berries are formed, which are edible and much favored by squirrels and chipmunks and even by the black bear, while the deer browse on the foliage. The north country may not have the showy dogwood, but we can rejoice in the beauty of this slender, graceful shrub as it blooms so briefly on the rocky hillsides, in the open woodlands, or bordering the swift brooks. It gladdens our hearts in the spring, feeds the deer in the summer, and provides berries for squirrels, chipmunks, and bear in the fall. Quite an accomplishment for a delicate little shrub! [6]

LILAC TIME

Now is the brief season of the lilac bush, modest and enduring symbol of the depth and permanence of New England traditions. It has given a name to color, perfume, poems, songs, story.

Translated into many languages, its name is upon the lips of millions in many lands. Yet it remains unspoiled by such widespread fame. It is still the sturdy, wholesome dooryard emblem of the New England home.

With what eager anticipation has it been planted at the threshold of new, bravely begun homes.

With what poignant grief has it been left behind for long bitter migrations from whose hardship and loneliness homesick thoughts have turned in anguished longing.

To what strange and distant homes have its roots been transplanted, there to grow blossoms and, in their turn, be abandoned again.

On this very day in mountain pastures and long deserted roads, over the graves of dead homes bloom the lilac bushes planted by the founders of those pioneer households. Many of those graves would be otherwise indistinguishable, their timbers long since buried, their cellar holes filled in and grassed over.

Were it not for the steadfast lilac bush, there would be nothing to mark that here once dwelt human souls who shared happiness, sorrow, hope, and despair.

Who lived there, whither they went or what their adventures, nobody knows. No descendants make annual pilgrimages to remember and decorate these forgotten graves of the homes of ancestors. But each year at this season, the lonely, faithful lilac bush blooms again and lavishes its sweetness in memory of the hands that planted it. [7]

LEAVING WILDLIFE YOUNGSTERS WILD

People do more harm to wildlife youngsters by picking them up and taking them home than they would if they left them alone, according to the Vermont Fish and Wildlife Department.

This is the time of year when fawns, ducklings, raccoon kits, and countless other young wildlife are trying their legs and wings for the first time. Some will be found and taken home by well-meaning people who do not understand the ways of nature. Sadly though, these animals almost always suffer in captivity and cannot readjust to living in the wild.

A little spotted fawn with big brown eyes and oversized eyelashes is hard to resist. A doe will purposely leave her newborn fawn in hiding, often in tall grass, and then go a short distance away to feed. Usually she will catch the scent of an approaching human and move away without being seen. The immediate reaction of a person stumbling onto the hidden fawn is to assume it is abandoned. Too often the fawn is carried home to "save it." The fawn soon gets used to its "protector" and periodic bottle feedings. It will never be a completely wild deer again. If it does grow up and is released, it will not be wary of humans.

Little raccoons also fall victim to people with good intentions. Their appealing curiosity and cuddly inclination make them irresistable. But they change when they grow up. Natural aggressiveness and an unpredictable disposition make a big raccoon dangerous. Children have been killed and seriously disfigured by pet raccoons.

Baby birds also are picked up by people who think they are abandoned. Fledglings are fed by their parents for several days after leaving the nest. The older birds have no trouble finding the hidden babies when they bring food, because the fledglings put up such a fuss at seeing their parents. Too often young birds kept in captivity get improper food and die. They fare better left alone.

Leave wild animals wild. Enjoy and observe them from a distance. Thinking of wild animals in human terms and upsetting their natural upbringing will do more harm than good in the long run. These are some of the primary reasons why it is against the law in many states to keep wildlife in captivity. [8]

DANDELIONS

Dandelion,
You'd make a dandy lion
With your fuzzy yellow ruff.
But when you're old
You're not so bold
You're gone with just one puff!

Dandelions may well be the first flowers that many of us learned to call by name; their abundance, their relatively early blooming, and their use or nuisance reputations bring them to the attention of children and adults alike. The name dandelion is also easy to remember, especially if you know its derivation, *dent de lion* or "tooth of a lion."

Dandelions are members of the Composite family, which gives them a number of famous relations: daisies, hawkweed, sunflowers, chicory, goldenrod, burdock, and aster, to name a few. The word composite describes flowerheads composed of many tiny individual flowers; when you pick one dandelion blossom, you actually pick approximately 300 dandelion flowers or florets.

During the first year of a dandelion's life, the leaves, which grow in a circular pattern parallel and close to the ground, are the only visible part of the plant. Busy all that first summer photosynthesizing to make food, which is stored in the long carrot-shaped root, the first year leaves help the plant get a headstart for the second and succeeding years in making flowers. One dandelion plant can produce many flower heads at a time (sixteen have been counted) and in rapid succession. When a dandelion flowerhead blossoms, the individual florets, each a complete miniature flower, mature in circular rows starting at the outside rim.

After studying dandelions, it's hard not to admire them. Humans, however, have very mixed feelings about the plant. On the positive side of the ledger, new spring dandelion greens are delicious and nutritious, the blossoms make tasty wine, and to some, green fields blanketed with yellow dandelions are one of May's most beautiful sights.

Those who prefer only grass in their lawns, and only planted vegetables in their gardens may well have a legitimate complaint against dandelions; they are hard to battle and nearly impossible to conquer. Why? One: The long central taproot won't be pulled up; it must be dug up. This is easier said than done because of its length and because of the tiny secondary roots that grow out from it and help anchor it to the soil. If any parts of the root are left behind, each root fragment can give rise to another entire plant. Two: Dandelions produce flowers and thus seeds from April into November (few actually blossom in midsummer). And three: Once formed, the seeds are great travelers, relatively unfussy about where they land because they can germinate in such a great variety of habitats. No wonder dandelions are so successful around the world from mountaintop to valley floor. [9]

DEW

Sometimes it's hard to know what to put on your feet on a June morning. You know the grass is going to dry but it's soaking wet first thing and who wants to wear boots all day or spend the morning slogging around in wet shoes? A solution is a pair of almost-too-big, easy-to-get-on-and-off, black rubbers.

There are two water sources that contribute to morning moisture. The first source is the air. Water vapor is always present to some degree, but how much the air can hold varies. The cooler the air, the less moisture it can hold. Fog or mist form when the air is so saturated that it can hold no more and the vapor condenses around tiny particles of dust or salt, into droplets of water. Rain falls when those droplets become too heavy to remain suspended. The colder it is, the sooner accumulated water vapor becomes fog, mist, rain, or dew.

In summertime, there is often a sharp drop in temperature at night, especially on a clear night, because of radiational cooling. Long-wave radiation that is given off by the earth's surface, unless obstructed by clouds or pollution, takes away much of the heat accumulated during the day. The moisture level in the air may not change at night, but the air's ability to hold that moisture decreases as the temperature drops. When the air temperature lowers to the point where the air can no longer hold moisture (that temperature is called the dew point), dew may form. Dew is moisture that condenses into droplets on cool surfaces like blades of grass, cars, or outdoor furniture. So your feet may get wet because of last night's rain, mist, fog, or dew.

Or they may get wet from water that comes out of the plants themselves during the night and accumulates on the tips of their leaves. This is called guttation. Rose, strawberry, and grass leaves are the best known examples in which guttation occurs. When there is abundant soil moisture, many plants transpire (emit water vapor) all night long. If the air cannot absorb that moisture because it's too humid, or because cooled temperatures have lessened its capacity to absorb, the transpired water vapor remains on the leaves and condenses into droplets. We dislike wet feet, but we should treasure morning's glistening grass.   [10]

HIKING

Summer is upon us — so fleeting a time is it that we must hurry to accomplish all those wonderful programs that we planned during the winter. "Let's have a family hiking trip," we had said in February. Accordingly, ten of us, aged between eight and sixty, each with backpacks of appropriate size and weight, set out the other day on a trail that winds over tops of mountains. The first day was a short hike, but long enough for legs and shoulders unused to carrying households on one's back like a turtle. Tents quickly went up at the camping spot like large colorful mushrooms, and over the hissing camp stove a delectable stew was soon bubbling. It's surprising how comfortably one sleeps on the ground, barring a root or two of course, and how short the nights are in early summer. Then a full day's hiking with a two thousand foot gain in elevation in torrid heat, all the while keeping blackfly oil close at hand. The best thing about blackflies is that on a short hike the bites don't itch until you're back home. The view from the mountaintop was cloudy, but we sat on the bare ledges nibbling our trail gorp and feeling that wonderful sense of satisfaction that all climbers experience whether it's Mount Everest that they've struggled up or only old Baldy. Gorp, by the way, (good old raisins and peanuts) is a trail snack of anything you have at hand but usually made of nuts, chocolate bits, and dried fruit.

Ominous black clouds in the west persuaded us to move on to the next camping site. We hurried down off the top, for mountain peaks and ridges are no place to be in a thunderstorm. "Mountain thumpers," as they are known, have been the cause of death to a number of hikers who have been killed by lightening in those high exposed places. We walked as carefully and as fast as we could under the canopy of the trees over the rough trail, but the thick black cloud turned the day into dusk. A few seconds before the gods unleashed all the lightening bolts they had on hand, accompanied by mountain-shaking thunder and sheets of rain, we reached the wooden shelter.

No tenting that night, so ten of us slept fitfully on wooden bunks while a large porcupine tried to gnaw through the door. On the last day we came off the mountains, muddy, a trifle bruised, flybitten, but very pleased with ourselves.   [11]

STARS ON A SUMMER NIGHT

On summer nights we like to sit outside in the cool evenings as the sky pales and the stars and constellations wheel across the heavens. There's the Big Dipper on its annual journey, endlessly circling the North Star, and there are the two pointer stars. The Big Dipper is the easiest constellation to find in the northern Hemisphere, and it was probably one of the first to be named by the ancients. There are lots of names for it; in the United States it's the Big Dipper, but the English call it the Great Bear or *Ursa Major*. It's also known as Charles's Wain. A *wain* is a wagon, and the Charles referred to is the Great Charles of the Holy Roman Empire, or Charlemagne as he is better known. The Greeks also thought of the Big Dipper as a vehicle. They thought it was a bier on which the dead were carried, and, in fact, the ancient Egyptians believed that the dead were carried to the sky neighborhood around the North Star, where stars are visible all year-round and therefore eternal. The Romans thought of the Big Dipper as the Helix, the twisting one, because it twists around the North Star.

But bear names are the most common throughout the world. Maybe the Great Bear was named many thousands of years ago, perhaps in the Ice Age, and the stars in the constellation were thought of as a divine animal with its neck stretched out. If you turn the tail into the neck, the Great Bear is like one of the polar bears with neck and head exaggerated and legs pointing inward that the Eskimos carve out of walrus ivory. As we lie back and look into the summer dark sky, we are following countless ancestors, those country folk, who by their huts and houses looked up, too, and made up stories of the constellations endlessly and silently wheeling over their heads.   [12]

THE SCYTHE

The hayfields have just been mowed for the second time. The mowing machine laid the grass in flat rows on a Friday, and by Sunday afternoon the hay had been carried off to be stored in the barn, a one-man operation. The tractor towed the baler, which in turn shot neat, rectangular bales up and into the attached hay wagon. Now the yellow fields will have to wait for the sun and rain to turn them green once more. But machines can't get into corners and along edges, where the weeds lurk to take over the fields. I must get out into the fields with a scythe.

First, I'll make sure the blade is razor-sharp by going over it with a whetstone. Scything with a dull blade is tiring and forces the mower to swing with a violent motion. Now I'll take up the long curved snath made of ash — the snath is the handle — with two grips or nibs. There's an easy rhythm to scything, both feet step forward, then swing, both feet forward and swing. The heel and toe of the blade must be horizontal, and just above the ground. As I start down the field, I think of the old-timers who would start out about four in the morning, and with two working together could get a couple of acres down before breakfast. They used to have scything contests where two men would start level, about ten feet apart, mow their way up the edge of the field, turn across the end and back down the farther side. If the inside man could mow the other out of his swath, by getting to the end of the field first so that the outside man couldn't turn to the right except by crossing in the inside man's swath, he had won the contest.

Not many people can use a scythe correctly now. If they try, they swing it like a golf club, slashing at the grass. I guess the scythe has almost come to the end of its useful life. It's a tool that has been used for countless thousands of years, ever since someone got tired of pulling grasses up by hand and used an animal jawbone instead. Then came a bronze blade with a short handle, and it was the Romans who first used a long handle so that the mower could stand upright.

I've been looking at a reproduction of a medieval painting of a man cutting hay, and his scythe looks just like mine except that the handle is straight and has only one grip. Well, scything may be getting to be a lost art, but I'm going to keep swinging this blade as long as I'm able. [13]

BRIGHT PASSAGE

Splashed among the reds, purples and rusts that bedeck the forest in autumn, the lemon yellows of birch leaves add a lighter touch. It's really a matter of chemistry and climate. But when those leaves begin to turn, we humans turn to metaphor for an explanation. It's nature's paintbrush at work, we say.

Botanists tell us the colors were there all the time. The autumn hues, they say, are really an unmasking of colors that were hidden by the darker green of chlorophyll, the substance that uses light, carbon dioxide, and water to fabricate food for the plant.

The changes of the season really begin in late summer, when the shorter days make less light available. Longer nights and lower temperatures lead to the breakdown of chlorophyll. At the same time, a hormone at the base of each leaf stem, or petiole, begins to form a circlet of cork, the abscission layer, that weakens the connection until leaf and tree part company.

Why, though, do birch leaves always turn yellow? Primarily because they contain an abundance of yellow pigment called xanthophyll. Colors in other trees come from chemicals with equally high-sounding names, like carotene (orange-yellow), anthocyanin (red), and tannin (brown).

The colors signify that the leaves are done with making food, but even after they flutter to the ground, they have yet another role to fulfill: they protect the exposed ground from erosion and their decaying bulk replenishes the soil. Life, like the rhythm of the seasons, goes on.

Soon, the wind and snow of winter will sweep across the forest. But always there will be the sure knowledge that spring, like the hidden colors of autumn leaves, is waiting in the wings. Even a botanist can't fault a feeling fostered by the heady autumn air that it's all part of a grand cadence, that after the quickened beat of spring and the lazy drone of summer, the frosty winds of change are welcome. [14]

THE MAPLE CASEBEARER

An autumn walk in the woods can reveal many secrets that were well kept or at least less obvious during the summer months. As feet shuffle through the dry golden brown carpet, or step silently across the damp mat, many details about the leaves become noticeable. Sugar maple leaves, especially, with their circular holes, tiny fingerlike bumps, and unevenly shaped spots, raise a number of questions. The reason for most of these physical peculiarities is simply insect invasion. But let's take a look at the most amazing of these invasions.

Have you ever noticed the nearly round, small holes in sugar maple leaves? And further, have you seen the same-sized dusty green or brown leaf discs on the forest floor? What causes them? Stop a moment and look the next time you're in a sugar maple wood. These discs are the late summer and winter housing of a very small and very common caterpillar called a maple casebearer, which will emerge in spring as a small, half inch, steel-blue moth. Why are the holes cut and how are the discs made?

In summer, the adult moth lays eggs on the surface of the maple leaf, and when one of the eggs hatches, the young larva bores into the middle of the leaf between the top and bottom layers. After a few days the little caterpillar makes oval-circular cuts in the upper and lower surfaces of the leaf, careful to make the lower section slightly larger than the upper one. It next sews the edges of the two sections together with silk. The larva, now enclosed in a roundish case, pushes the front part of its body out and drags the case onto the surface of the leaf, at the same time flipping the case over so that the larger section is on top. It looks somewhat like a miniature turtle as it drags its house around.

The larva moves slowly across the leaf, eating and growing as it goes. When it finds life a bit cramped in its disc-house, it merely sews its present house to a slightly larger oval area of leaf, and then cuts a slit around the enlarged oval. If it elects to stay on top of the leaf it once again has to flip its house over so that the larger leaf section is on top, or, without flipping, it can merely drag the case onto the lower surface of the leaf. It is quite an admirable feat to cut the oval disc out of the leaf while sitting in the middle and managing to keep it from falling to the ground. In autumn, the maple leaf falls and there the well-protected pupa, snug in its many-layered leaf house, spends the winter.

It would be hard to devise a more ingenious housing plan. [15]

PONDS CHANGE IN AUTUMN

In the chilling winds of early winter, shriveled leaves are blown over the pond's surface like tiny boats. A walk to a nearby pond often provokes fond memories of the past summer—the almost audible sounds of children splashing and screaming as they cooled off during one of our few hot days.

What secrets are hidden in those limpid pools? While much of the life on land is being prepared for a winter's nap, the pond is about to experience a rejuvenation of sorts.

As the sun sweeps low over the horizon and winter approaches, the pond cools down. From their summer high temperatures in the seventies, the ponds slowly begin to approach the point of freezing. Initially the heavier, colder water sinks to the bottom, pushing the warmer water to the top. When this happens, a churning motion begins, and the pond is stirred up. Dead leaves, silt, and tiny animals lying on the pond bottom are carried aloft in this current and distributed throughout the depths. Fish, frogs, aquatic insects, plants, and other forms of pond life make food of these nutrients from the deep. Fall is a time of enrichment for pond life.

After this last spurt of energy and activity, many pond creatures, such as turtles, frogs, and salamanders, will burrow into the mud to hibernate. Although the growth and movement of other pond life slows down during the coldest months, many insects and fish will remain active all winter beneath the ice.

But a question arises — if colder water becomes heavier and sinks, why doesn't a pond freeze from the bottom up instead of from the top down? The answer is a bit complex. As water cools, it becomes heavier and denser until it reaches just over thirty-nine degrees F (four degrees C). After this point, it once again becomes less dense and begins to rise. Water at the freezing temperature (thirty two degrees F or zero degrees C) is less heavy than slightly warmer water and is thus buoyed by the warmer layer below. If ponds did freeze from the bottom up, most pond life would perish. [16]

BANKING THE HOUSE

In those days when meadow grass was crisp with frost every morning, but before the ground had frozen, we would set aside a day for banking the house. It was a ritual without which a family would face the winter months with some trepidation.

Work on the farm would stop following morning chores, horses were hitched to the wagon, and boards hauled from shed to lawn. With iron bar and sledgehammer, three-foot stakes would be driven in a row about eighteen inches from the house's foundation.

Boards were nailed to the stakes to make a neat trough into which leaves were dumped for insulation that kept cellar and floors at a bearable temperature all winter. It was energy conservation long before the term was coined.

The wagonbed, with high sideboards, was piled high with leaves—treaded by the youngest and least useful of the youngsters. Leaves, if they aren't wet, are fluffy and need a good bit of treading when dumped into the wagon for hauling, as they do when thrown into the trough for banking.

We remember what seemed like endless hours of running back and forth on leaves—wallowing in them up to the waist and savoring their fragrance as they were crushed and jammed into place to forge a tight seal against the frost.

Unlike modern homes, the farmhouse was apt to have a foundation of stones, and often the foundation was far from airtight. It was important to have the banking reach above the bottom of the clapboards—compacted enough so it wouldn't settle under the weight of snow.

Canned goods, syrup, apples, potatoes, and the ripening cider would survive the winter in a delicious, musty coolness when the ritual was observed so strictly a field mouse would have trouble working his way into the boy-packed leaves.

It was a day set aside for protecting the harvest from all that winter promised, and one of the more satisfying workdays farm life could offer. [17]

WINTER BIRD FEEDING

Most of the birds have left for points south by now, but there are several species that spend their winter in northern New England. This is the best time of year to observe these residents, as the trees and bushes no longer offer concealing foliage, and the interested observer also can attract many species to easy viewing distance by putting out birdseed. A bird feeder can provide hours of winter viewing pleasure, and now is the time to start feeding!

There are several things to be considered before one begins a bird feeding program. First and foremost, once a feeding program is begun it must be continued without fail until spring. If you go on vacation for even a weekend and neglect to fill your feeders, some birds who have grown to depend on your feeder as a food source may starve to death. The feeders should be located in an area where there are shrubs or trees close by, so the birds will have adequate protection from the elements and predators. The feeders themselves can be homemade or bought, just as long as they hold seed. It's helpful if the feeders are covered, so the seed won't be buried by snow, and also if there are small drainage holes in the bottom of the feeder so rain or meltwater won't collect in the bottom and cause the seed to spoil. There are feeders designed to be squirrel-proof, for these furry creatures can consume much more seed than the birds at a single sitting, but don't count on squirrels being deterred from them. They'll find a way.

As for what to feed the birds, the choice is up to the individual. The standard wild birdseed mix sold in grocery and other stores is excellent for attracting a wide variety of birds, including tree sparrows, cardinals, purple finches, and blue jays. Sunflower seed is relished by chickadees, evening grosbeaks, and red-breasted and white-breasted nuthatches, and it can be purchased separately from the mixed seed, although some is included in the mix. The more discriminating bird-watcher may want to fill a feeder exclusively with thistle seed, which is extremely attractive to wintering goldfinches, pine siskins, and common redpolls. Cakes of beef suet will attract woodpeckers, as they use the suet as a substitute for the insects that they normally eat. A single feeder will provide hours of enjoyment for the human observer and be very much appreciated by the feathered winter residents. [18]

SNOW IS NATURE'S SECURITY BLANKET

When winter snow lies deep in the woodlands and all but buries field and briar patch, the outdoorsman casts an anxious eye outside. He knows malnourished deer can starve in their rutted yards in winter, and pheasants can freeze when blizzards strike.

But nature is never self-defeating. The role of snow in the ecology of northern forests and farmlands is a complex one and, on balance, it is vastly more beneficial than harmful.

Snow not only benefits wild creatures directly but it also in less obvious ways affects the soil, water, and plants upon which they depend. The fish in many lowland streams would not exist were it not for meltwater from slowly shrinking snowfields. Countless tiny plants above timberline can survive only if protected from winter's drying blasts by a layer of snow. Many plants would die, their roots torn and exposed by frost-heaved soil, were it not for snow's moderating influence upon the temperatures beneath it.

Insulation is one of snow's most valuable gifts, massed snowflakes trapping dead air in countless spaces to form one of nature's most efficient insulators. In extremely cold weather, the temperature of the ground beneath a deep blanket of snow may be fifty degrees warmer than the air above it.

This protection is of inestimable value to wildlife. With the approach of winter, the forest floor is populated by a dense concentration of hibernating creatures. Queen bumblebees, woolly bear caterpillars, wood frogs, spring peepers, red efts, crickets, and countless other forms of wildlife snuggle beneath the forest duff and leaf mold. Rotting logs are occupied by snails, wireworms, spiders, centipedes, and hornets. Box turtles, toads, snakes, and beetle larvae burrow underground. Grass clumps harbor spiders and stink bugs; stones hide crickets and pupating larvae.

Among the less hardy, the higher temperatures induced by snow cover can mean the difference between life and death. A recent study showed nearly 70 percent of certain larvae hibernating in trees perished during the winter, while those wintering beneath the snow lost less than 10 percent of their numbers. [19]

SNOWSHOES

I have come up to our spring that feeds water by gravity to our house. A heavy rain in between our cold spells washed some mud over the strainer at the bottom of the spring so that we weren't getting much of a flow. I cleaned it out so we should get a good run now. As far as I know, this spring has been here since 1790 when the house was built, and I wonder how many times the old-timers came up to do just what I have done.

I had to get up here with snowshoes, or I would have had a tough time making it in this deep snow. From what I have read, snowshoes were in use over 2,000 years ago. Probably they came to North America by way of the land bridge that could have been between Siberia and Alaska. We know that native Americans depended on snowshoes to get around in the very early days of our history. Today, trappers, game wardens, and woodsmen can't be without them.

There is much talk about cross-country skiing these days, but I think snowshoeing has been overlooked. As far as cost goes, there isn't much difference in fitting out between the two. A beginner can get along very well after an hour or so of practice on snowshoes, while it will take a lot longer to master long lightweight cross-country skis.

Snowshoes come in various sizes and shapes. The stubby bear paw is designed for travel in thick cover and climbing, and we use them collecting sap in our sugarbush. The long Alaskan Trail shoe is best for moving through relatively open country on loose powder snow.

There are many types of snowshoe bindings, but the average style, such as I am wearing, has a toe strap of wide leather into which the boot is inserted, a strap that goes around the heel and another that goes under the foot, over the instep, and under the arch.

I think my rubber-bottomed leather-top boots with removable felt linings are a necessity for good snowshoeing. It's about zero right now, I've been out about an hour, and my feet are nice and warm.

The webbing on the snowshoes that keeps me on the surface is made of rawhide lacings. If I should leave them up in our sugarbush, the family of porcupines there would do a good job of chewing up the rawhide. I make sure to hang the snowshoes on the ceiling of the toolshed.   [20]

ANIMALS KEEPING WARM

Did you ever wonder what happens to animals during the cold and blizzards of winter? They must keep from freezing, just as we do when we are outdoors.

Many mammals are capable of growing thicker fur in winter. Dense woolly hairs help to trap body heat in thousands of tiny air spaces. Of the fur of several animals, including the red fox, black bear, wolverine, red squirrel, hare, wolf, and muskrat; the wolf's is the warmest. The woodchuck's body temperature drops to thirty-seven degrees F and his heart beats only four or five times per minute; he and the little brown bat and jumping mouse are true hibernators. Raccoons and squirrels are active most of the winter but sleep during bad weather. Chipmunks spend the winter underground, sleeping and munching on food stored in interconnected storage chambers during the fall. Renovated bird nests are the winter homes of many mice, who add a roof of shredded leaves and bark, and cattail fluff insulation. Beavers also are active during the winter, but they are rarely seen, locked under the ice of their ponds. They breathe through air holes or from bubbles under the ice.

Birds have the ability to fly south, to where the winters are milder and they are able to find food. Those that do stay can fluff up their feathers for more insulation and find a good stout tree for shelter from wind and snow.

The ruffed grouse has a unique solution to the problem of keeping warm during especially cold or blizzardy weather. If you find birdprints leading away from a hole in a snowbank, you may have discovered the shelter of a grouse. Experiments have shown that unpacked snow is a quite effective insulator. At an air temperature of minus twenty-seven degrees F, the temperature under seven inches of snow is plus twenty-four degrees, a difference of fifty-one degrees. At minus thirty-two degrees, the temperature under twelve inches of snow was found to be plus thirty-one degrees. What makes this efficiency are the tiny pockets of air trapped between the flakes of snow, which almost eliminate the passage of heat.

Snow is two to three times more efficient as an insulator than loose sand. The snow blanket also prevents the winter cold from killing many plants and provides water for new growth when it melts in the spring. [21]

WINTER FEET

Before ice and snow arrive in the Northeast, we put on winter tires. Cars with winter "feet," studded or deeply grooved snow tires, move better than those with summer "feet."

Some animals too have feet adapted for winter travel. The ruffed grouse grows horny comblike fringes on the sides of its toes in the winter. Extra hair grown between and around the toes increases the size of the snowshoe hare's foot, and the Canada lynx grows extra fur on its broad flat feet. Feet with extra surface area tend to sink less into the snow because there is less weight per square inch.

Animals don't have to grow extra combs or hair to increase the surface area of their feet. Members of the dog family, for instance—foxes, coyotes, dogs—spread their toes and put the whole foot down at once to help distribute weight. Watch a large dog walking on breakable crust—it places each foot gingerly and flat on the snow and gently shifts its weight from one leg to the next.

Just as snow tire surfaces are designed to grip, so are the pads on the feet of many creatures. These adaptations were probably not to improve winter travel capability, but they certainly help on slippery surfaces. Squirrels not only have bumpy pads on their toes and on the palms of their feet, but they also have needle-sharp toes to dig into bark or ice or crust. A porcupine has nonskid pads on its feet as well as modified spines on the underside of its tail.

Horny pads within the sharp toenail rims of deer hooves help hold onto crust; on ice, however, a fallen deer is almost helpless as the skidding toenail rim makes getting up nearly impossible.

Winter travel is a challenge for all creatures who must move about. Some animal feet are examples of excellent devices to meet the challenge, and humans have copied many of them. [22]

SNOW FLEAS

In late winter, you may have noticed large numbers of minute, wingless insects jumping about on the snow surface. Commonly known as snow fleas, these insects belong to the order, *Collembola* or springtails.

There are some 2,000 species of springtails in the world, and they can be found on every continent including Antarctica. They occur in moist vegetation, forest litter, under bark, in decaying wood, and on water surfaces. They belong to one of the earth's most beneficial janitorial unions, the scavengers, feeding on decaying matter such as leaf litter, thereby playing an important role in the formation of humus.

Most species of springtails possess a long slender tail-like structure, or furcula, at the posterior end of the abdomen. The tip of the furcula is held close to the undersurface of the abdomen by a clasplike structure. When the clasp is released, the furcula snaps against the ground and propels the insect forward and upward.

Research seems to indicate that snow fleas and other springtails swarm in response to population pressures. Under ideal conditions of food and temperature and when predators are scarce, the population rapidly increases, creating a food shortage. On a suitable day, when temperature and humidity are right, the insects migrate in great numbers to the surface of the snow, emerging around the tree trunks, rocks, and weed stems where snow has partially melted away. Great numbers of them will die within twenty-four hours, while a much smaller number will find their way back down to the leaf litter to survive the rest of the winter. There is still much to be learned about springtails before we can be sure of the reasons for the late winter appearance of these insects. [23]

A LONG WINTER'S NAP

When winter approaches with its blustery winds, subzero temperatures, and seemingly endless wastelands of ice and snow, the creatures, so contented with summer's ease of life, have to be prepared for the upcoming hardships. Although cold and storms make life more difficult, probably the greatest challenge in winter is finding food. The rich greenery is gone, as is the once abundant supply of small creatures.

Different species handle the challenges of winter in various ways. Many birds and some bats migrate to warmer environments, where food is plentiful. A number of seed-eating songbirds, carnivorous animals, and many herbivores such as voles, deer mice, and white-footed mice remain active in the cold, spending almost all their waking hours in search of a meal. The supply of food is the critical factor. The Arctic fox can maintain its body temperature when the atmospheric temperature is minus 112 degrees F and, like the snowy owl, handles a frigid climate beautifully as long as there is a sufficient amount to eat. The snowy owl migrates south during the winters if its food supply is limited.

When faced with the difficulty of finding food, the shorter days for this search, and the bitter cold, many animals retire into snug homes and go into a deep sleep. Opossums, black bears, striped skunks, raccoons, and Eastern chipmunks all spend long periods in a sound sleep, waking occasionally, usually on a warm day, to look for something to eat. The chipmunk puts by a supply of nuts and seeds in its burrow to help get through the winter. The other deep sleepers are omnivores and will venture out of their dens or burrows to seek food.

The final way animals cope with winter is to hibernate, quite a different behavior from the deep sleep of bears and skunks. Unlike the sleepers, whose vital signs do not change significantly, hibernators experience remarkable physiological changes. Body temperatures, breathing rate, and endocrine controls all fall off noticeably. Studies done on hibernating woodchucks show their body temperature drops from around ninety-seven degrees to thirty-eight degrees F, their pulse goes from one hundred beats per minutes to five per minute, and their rates of respiration fall from as much as one hundred respirations per minute to one every four or five minutes. [24]

"HABITAT" GARDENING

When February rolls around during a particularly harsh winter, our thoughts often turn to the difficult conditions for wildlife survival. As a means of easing our minds, we also think about the coming spring and get an optimistic feeling considering what crops we'll plant at that time, and how life in general will be a little easier for all. When you plan your gardens, you might look out over that backyard (or front yard, for that matter) and consider how, through plantings and a little rethinking, you can make your yard more inviting and productive for wildlife.

Habitat is a word that conjures up visions of expansive wilderness areas, pristine preserves, and great quantities of land. Basically, habitat is as simple to understand as the word *home*; it includes all the essentials for survival—food, water, minerals, thermal cover, escape cover (from predators), reproductive cover, and nesting cover. For an insect, habitat can be as small in size as the dead tree it lives on. For a white-tailed deer, habitat size must be considerable to meet all its needs.

You may consider your influence on the components of an animal's habitat to be minimal at best—especially in your backyard. But there are a few simple things an average homeowner can do. First is to consider the fact that a variety of plantings is most conducive to attracting a greater variety of animals and birds. A yard with large trees (both pines and broad-leaved), with shrubs and bushes (such as dogwoods, brambles, and honeysuckles), and a grassy field area will invite a much greater variety of songbirds, small mammals, and larger wildlife than a pure plantation would. With a little research, you can find out what plants attract what wildlife, and which will do well in your area. For example, if you're interested in attracting songbirds, red cedar, honeysuckles, highbush blueberries, and brambles not only offer variety in size, shape, and color for the aesthetics of landscaping, but also offer a feast for all the colorful, wild visitors you may want. Deer and cottontail rabbits love a field of alfalfa, clovers, fescue, and crown vetch, especially if it's near the edge of a wooded or brushy area where they can hide.

So on a blustery February day, take the time to carefully look at your yard. Use your imagination, and envision the feast and protection you could provide wildlife. You'll also find you'll be adding a tremendous amount of beauty to your own habitat. [25]

Plants & Gardens

HOW SOILS ARE MADE

Ancient soils existed before the last Ice Age. Then a huge ice mass, nearly a mile high, moved across the land, scouring off these ancient soils along with much of the rocky material beneath. Together they were ground into a complex mass and carried varying distances. Eventually, as warmer weather melted the ice, they were dropped. This hodgepodge of rock and soil containing various proportions of gravel, sand, silt, and clay is common much of the country and is called glacial till.

Now try to imagine the tremendous amount of water liberated as the ice melted. This gouged deep valleys, created many lakes, and shifted soil materials from place to place. Glacial waters sorted materials of different sizes just as present-day streams do, only on a giant scale. These materials eventually were deposited as gravel, sand, silt, or clay. Large particles were carried short distances, but clay, the finest, was carried to quiet lake waters before being dropped. Present-day gravel pits, sandy areas, and clay beds are the evidences of such activity.

Lack of vegetation and high winds, after the water disappeared, also account for deposits of sand and silt in many places. From such parent materials, northeastern soils have originated. Other destructive forces, including freezing and thawing, have continued to break rocks down into soil.

Coupled with these forces have been soil building processes that make soils differ in other respects. Variations in temperature, rainfall, kind of plant growth, and the length of time the soil has been developing all play a part. Thus soils differ in depth, size of particles, supply of nutrients, amount of pore space, degree of wetness, humus content, and color.

Most soils exhibit characteristic layers called horizons, layers that have resulted from the interplay of all these forces. River bottom soils, many having been deposited within the memory of man, are too young to develop a profile of horizons. It takes at least a hundred years for soil building to even get started.

Because the glacier departed at least ten thousand years ago, most New England soils are quite old. During this period they have lost much of their natural fertility by vast amounts of rain water running through them. As a result, they generally are acid and low in fertility compared to soils of the Midwest. Still, differences in inherent fertility soils affect the crops one can grow.

Perhaps you can now see how it is possible to have hundreds of kinds of soils, no two exactly alike. The Soil Conservation Service, to simplify soil study, has divided soils into 150 types and given each a name. [1]

ROADSIDE FLOWERS

How lucky we are to have flowers, especially those that are quick to grow in waste areas. Among the many kinds to bring color to our roadsides and parking lots are two that on first impression could easily be confused. Perhaps you will be able to guess what they are from their descriptions. Both are yellow, both bloom from June through September, both have irregular lipped flowers, and both range in height from approximately six inches to two feet high. There the similarities end, and a closer look is necessary.

The first and probably more common of the two is bird's-foot trefoil, a plant often seeded by highway departments on road banks and on the edges of highways to help hold the soil and hide scars of recent roadbuilding. The flowers are bright yellow, quite small, perhaps a half inch in diameter, and grow in clusters of three to six per head. The bird's foot part of the name comes from the three to six seed pods that develop from the flowers and look like the curved toes of a bird's foot. The name trefoil refers to the three leaves that form the cloverlike end of the pinnate leaf. Actually there are two additional basal leaves, making five in all. Look closely at one of these rounded, lipped flowers and you will see they resemble their other cousins in the pea or legume family.

The second flower, again found on roadsides and in waste places, also has an unusual and very descriptive name: butter-and-eggs. This name refers to the mixture of yellow spurs and orange palate on each inch-long snapdragonlike flower. The flowers grow vertically in dense clublike spikes or racemes at the top of the plant quite different from the more horizontal clusters of bird's-foot trefoil flowers. The leaves also are very different, long and narrow, single, and rather stiff in appearance. This plant, unlike the native bird's-foot trefoil, is an alien said to be imported from Europe because it was used for colonial skin lotions.

When you next spot a clump of yellow flowers beside the road that you know are neither dandelions nor goldenrod, stop and check for bird's-foot trefoil or butter-and-eggs. Will you know which is which? Bird's-foot trefoil flowers: all yellow in a semicircular cluster. Butter-and-eggs flowers: orange and yellow in a spike. [2]

WEEDS: FOREIGN FRIENDS OR ALIEN ENEMIES?

It's hard to keep up with spring and early summer! Plants sprout, leaf out, and flower almost before one has time to notice the stages of growth. Many birds return, build nests, and even hatch their young almost before you catch a glimpse of them. There is a sense of urgency among plants and animals alike as each strives to take advantage of long days and warming temperatures. Summer is short, and survival of the species is at stake.

There are some kinds of plants that seem to feel an extra urgency. These grow overnight, as it were, in gardens, on roadsides, anywhere they find a fraction of untended, open earth; and the speed and tenacity with which they take over has earned them the name of weeds. Before dismissing these plants summarily as unwanted vegetation, it is only fair to consider their history, their amazing adaptations, and even their usefulness.

The history of weeds has only recently come under investigation; most concern about them has centered around their eradication. In a fascinating book, *Plants, Man and Life,* Edgar Anderson writes that weeds hve developed and flourished because of man. Where humans have settled even temporarily and created dump heaps or opened up the land, weeds, along with the earliest crop plants, found a suitable niche. "The history of weeds is the history of man." A list of weeds and their origins would support this theory. Very few of our nuisance plants are native; most were transported here by European settlers, either on purpose as food or ornamental plants, or by accident among crop seeds or belongings. Plantain, thistle, pigweed, burdock, crabgrass, dandelions, and stinging nettles are but a few of our alien weeds. A summer field is full of European imports—daisies, yarrow, buttercups—weeds to the farmer who needs good hay, flowers to those who merely look and enjoy. Settlers were guilty on two counts of colonizing the countryside with alien plants. First they brought the plants or seeds. But more important, they created a favorable habitat wherein these plants could flourish.

What kind of habitat is favorable to weeds? Your garden, for one; because competing vegetation has been turned over and the soil loosened, there is plenty of sun, and the area is well drained. Weeds, however, can survive in less luxurious surroundings and can in fact stand exposure to extremes of temperature, sunlight, wind, and rain intolerable to other plants.

The adaptations that enable them to survive harsh environments deserve our admiration for their effectiveness and efficiency. First, weeds are prolific; millions of seeds blow or are carried to all corners of the earth. Once landed, many can delay germination, for years if necessary, until conditions are suitable. The plants themselves, once sprouted, have myriad defenses to ensure survival in adverse circumstances. Long taproots and fuzzy or hairy leaves and stems defy drought conditions by finding hidden moisture and/or retaining it. Thorns, prickles, burrs, and stingers deter creatures, human and otherwise, who might pick, eat, or cut the struggling plants. [3]

POISONOUS PLANTS

Historically, poison from plants frequently was used to kill enemies, especially in the Middle Ages. In fact, the word *toxic* means *arrow poison*. Socrates chose to drink a liquid containing hemlock rather than be expelled from Athens. In earlier days people were more aware of the plants around them, but as other methods of killing became more popular, this common knowledge of poisonous plants declined. The recent enthusiasm for wild food foraging is causing trouble. Many people don't even know exactly what they are eating and consequently get very sick or even die. Guide book descriptions aren't always foolproof. Plants may change in form (and toxicity) as they grow, so pictures and text may be confusing.

Not all parts of a plant are equally toxic. Poison tends to concentrate in stems, seeds, and tubers. Alkaloids, for example, attack the nervous system, with the major symptom being a quite rapid lowering of blood pressure. Both bracken ferns and horsetails contain the protein compound of thiaminase, which causes death by promoting the breakdown of vitamin B. Mistletoe and sweet pea have toxic amines that can cause either paralysis or convulsions. The most common toxic substances in plants, however, are oxalates. These combine with calcium to form calcium oxidate, which is nearly insoluble and collects in the kidneys, causing malfunction and eventually death as the poison accumulates. Rhubarb, skunk cabbage, jack-in-the-pulpit, and philodendron all contain this poison.

What can you do to avoid being poisoned? First and foremost, never eat *any* plant you haven't identified. Birds, squirrels, and other animals can often eat with impunity berries and things that would greatly affect a human. Some people may be less affected than others. There is no really safe test for distinguishing an edible plant from a poisonous one, and all the major classes of plants contain poisonous members. Children should be taught to recognize poison ivy and other skin irritants and be careful of what plants or twigs they play with or use as skewers for cookouts.

If poison is suspected, call your doctor immediately. Do nothing until you obtain your physician's advice. Various poisons react differently to one's body chemistry and may need special treatment. [4]

CREEPING MENACE

What grows a mile a minute, is able to leap up an empty house in a single summer, and is nearly indestructible? It's kudzu!

Kudzu? That's right. Kudzu is an exotic vine with a broad three-pointed leaf and woody stem, which was imported to the United States from Asia. And while many southerners call it the mile-a-minute vine, it doesn't grow quite that fast. Actually, at most, a stem can grow about one foot per day.

Once kudzu starts to grow, it doesn't want to stop. The vine has engulfed much of the South, where long growing seasons and abundant precipitation are to its liking. Now it's spreading northward into Kentucky, Virginia, and Maryland, and westward into Texas and Oklahoma.

The plant begins growing in the early spring with its green tendrils radiating from its tap roots. It produces great quantities of foliage and, by late summer, clusters of fragrant purple flowers. Its tendrils can grow sixty feet in a season, often climbing vertical obstacles as high as forty feet.

During the Great Depression, it was discovered that the vine's deep roots, dense foliage, and rapid growth, along with its contribution of nitrogen to the soil, provided ground cover to control erosion, stabilize road banks, and rejuvenate nitrogen-deficient soil. Some southerners called it the "miracle vine."

Asians have always put kudzu to good use. In Japan the vines are used to make cloth, baskets, and paper, and hay is made from the leaves. The Chinese grind up the vines to make a popular kind of flour.

For most Americans, however, kudzu has lost its charm. Farmers have found that as forage, it is easily overgrazed, and much of the vine is woody stem, useless as hay. It also invades pastures and crowds out crops. An estimated one million acres or more of southern farm, forest, and pasture land are now covered by kudzu.

The vine isn't popular with foresters, either, as it engulfs all vegetation in its path, killing both large trees and saplings. Telephone companies don't like the way it sometimes pulls down their poles with its heavy, grasping vines.

So kudzu's aggressive growing behavior outweighs the positive qualities that caused many southern communities to form kudzu clubs and elect kudzu queens back in the forties. Now the United States Department of Agriculture lists the vine as a common weed. [5]

ANCESTRAL DIET

At a time when many are becoming more and more concerned about diet and whether to eat more protein or less, or more vegetables or more grains, it's interesting to know what our ancestors were eating. We do know about some of our remote ancestors, amazing as it may seem, and exactly what they ate. These forefathers lived during the Iron Age, at the time of Christ. How do we know? Because scientists have been able to dissect their bodies and determine exactly what the last meal was that they ate before they died. No, these are not mummies artificially preserved; they are bodies exactly in the same condition that they were at the moment of death. This occurred because these men and women were buried immediately in peat bogs, where the only change on the body was that of tanning. And, of course, tanning is a process of preservation.

What these Northern Europeans (for these peat burials occur mostly in Denmark) were eating the day they died was a sort of gruel made of grains. Scientists found the seeds of bindweed, dock, camomile, barley, linseed, and knotweed. In only a few burials was any trace of meat found. Does this mean that animals were too scarce to provide much food in those days? Or was this a special meal given to these people before they died? No one really knows whether these unfortunates were criminals or sacrificial victims killed to insure a better harvest. In any case, whether from necessity or choice, the ancient Danes of two thousand years ago seemed to have been on a mainly vegetarian diet. We have seen, in Denmark, the man who was dug out of the Silkeborg bog, and his perfectly preserved but very dark, tanned face seemed to have a peaceful quiet smile. Perhaps he felt his last meal of weed seed gruel had been a good one. A British archaeologist in an experiment, ate a meal made up of the same weed seeds and said it tasted terrible to him. But then, I suppose there's no accounting for taste. [6]

GARIBALDI'S GARDEN

We look at our vegetable garden out behind the kitchen with an appraising eye. The fence finally seems to have foiled the woodchuck family that lives under the horse barn, for the young cabbages appear thus far to be intact. In spite of a lack of rain the lettuces are flourishing. With their shades of green and red they are almost as colorful as the flower garden. The peas will be ready soon. Yes, right now, the garden world seems to be coming along as it should. But a shadow of discontent remains, for we remember the perfection of the gardens we saw in Italy, in particular, the garden still maintained at General Garibaldi's island home.

Garibaldi, the great Italian general and liberator, lived on a small isle a mile or so off the large island of Sardinia. He lived out his days there after uniting Italy, dying in 1882. Tucked in a corner off the low, white, rambling stucco house is the vegetable garden—a garden such as everyone has in Italy, the land of gardeners. Perhaps it was the setting on the beautiful little island that made us remember it, but what particularly impressed us was the raised beds. Now everyone is talking about raised beds, for they are said to produce more vegetables in a given space and, with the massed vegetable foliage, to retain moisture and block out weeds.

The other thing that has remained in our minds was the sheer artistry of the planting. We don't know if the gardener was paying attention to companion planting, but there were the majestic carchoufi (we call them artichokes), the eggplants, peppers, tomatoes, all vegetables dear to Italian cuisine. The young peas, carrots, and lettuces were shaded in the hot Mediterranean sun by an artfully pruned grape vine climbing on a trellis overhead. The effect of different greens and different shapes and sizes of the vegetables was as pleasing to look at as the most sophisticated flower garden. Well, our Vermont garden appears prim and ordinary perhaps, but it still looks pretty good to us. [7]

HIGH-YIELD VEGETABLES

"We have a small plot of land available for growing vegetables. It is in full sun. Which vegetables will give us the greatest return?"

Vegetable crop value is judged one of two ways: by price per pound of yield or by amount of yield per square foot of growing space. Some vegetables, cauliflower for example, are high-priced in the market. Thus, although it is a low-yield crop on a square-foot basis, its high price increases its value. Vegetables such as radish and leaf lettuce, which yield crops in a short time and in a limited area, are high producers. Also, they are harvested early, so the area they occupy temporarily can be replanted to other crops, which increases the yield value per square foot of space.

Green peas, sweet corn, brussels sprouts, and vining crops like melons, squash, pumpkin, and cucumbers tie up the garden area for long periods. They are low yielders and are not recommended for small gardens. The exception in this group is summer squash. Each plant requires at least four square feet of space, but fruit production is often overwhelming.

The tomato, when grown upright, is the most valuable of the vegetables. Crop yield is high, and early interplanting with short-term crops adds to the yield value per square foot.

Broccoli, after producing one good-sized head per plant, produces smaller side shoots all season, which are just as good to eat. Beans, both pole and bush, are high yielders. Again, the square-foot value is increased by preplanting with a short-term crop and by replanting after harvest.

You will achieve the greatest return if you plan carefully; garden intensively, leaving no soil unproductive; refertilize between crops; provide plenty of water and sun; and judiciously control insects and disease. [8]

HOW TO MAKE BETTER COMPOST

Making good compost isn't really difficult. Although some gardeners have had disappointing results with homemade compost piles, you won't if you follow a few simple rules.

It is possible to make compost in cold weather, but a little warmth does wonders in getting your compost pile started. When the weather has warmed up it's an ideal time to start a compost pile.

Before you begin, collect plenty of good organic waste materials. You'll also need a source of water and a simple enclosure to help you organize your pile.

Put a one- to two-inch layer of straw, hay, or other coarse material on the bottom, then gradually alternate layers of organic matter—grass clippings, lawn rakings, dried leaves, weeds, or old sod—with generous amounts of manure.

Dried cow manure is ideal, although a few handfuls of alfalfa-, cottonseed-, bone- or bloodmeal work just as well. Dog food and any other high-nitrogen organic plant food are also good substitutes.

The high-protein meal adds nitrogen to the compost pile in a form readily available to the beneficial microorganisms that break down and digest the raw organic matter.

A dusting of wood ashes or ground limestone also should be added once in a while. Or apply some rock phosphate, or a handful or two of good rich garden loam to increase the nutrient level.

After adding each new layer of organic material, generously sprinkle water on the pile. When the enclosure is full, pack the matter down well. Then level off the pile, making the top curve slightly inward to collect rain water. This will keep the pile continuously moist.

Now if all goes well, your pile should begin to heat up as extra energy is given off by the natural compost-making organisms. You can feel this warmth by digging into the pile with your hand.

When the pile cools down, turn the compost over with a pitchfork or spading fork to help make the partially decomposed raw organic matter into better humus. [9]

MANURE MEANS MONEY

Manure represents money—it is a farm product—just as surely but perhaps less directly than the check a farmer gets for milk. At current prices, a milking cow can produce nutrients worth $100. per year, an item not to be sniffed at, if you'll pardon the pun. Or think of each ton as five dollar bills.

Most farms, and Vermont as a state, import more plant nutrients in the feed bag than as chemicals. Grain becomes manure and then a fertilizer. Cows and their milk remove only part of it. Thus, most feed becomes fertilizer in the end.

Manure represents not only the addition of plant nutrients from some Western farms but, given a good legume forage program, it also adds large amounts of nitrogen to crops such as corn and grass. A few farmers have perfected their legume forage production to such an extent that no imported protein is needed. Their protein, and fertilizer nitrogen, come out of the air for free.

The waste of manure's value by improper handling can cause half or more of its original content to disappear. This means lost fertilizer, which either reduces crop yields or causes a farmer to buy extra chemical materials to maintain high yields.

While one can't generalize about manure's worth compared to taxes, it might be possible that farmers taking better care of manure can save enough money to pay their taxes. With $.10 nitrogen before 1973, the farmer might have suffered nutrient losses without being hurt. With $.25 nitrogen, this is no longer true. What about the future? Rumors already speak about $.35 and $.40 or more per pound. Future prices depend heavily on the world energy supply.

In addition, manure is an important source of potash for the nonleguminous crops, such as corn. Compared to nitrogen, potash is cheap—only about $.10 to $.15 per pound. But as long as we are dependent on another country (Canada) for a supply, future costs and availability are also vulnerable.

We hear a lot about self-sufficiency these days and organic food production. Both are unlikely to happen on large dairy farms but, given economic change, manure does offer the potential of being a better moneysaver and moneymaker than it is today. [10]

WITCH HAZEL

Once the aster and the goldenrod blossoms fade, and leaves start their journey to the ground, one puts away all hope of spotting the vivid colors of autumn during a walk along a hedgerow or woodland edge. All color does not vanish with the brilliant fall foliage, however. After its leaves have fallen, just when the woods are at their most stark, witch hazel bursts forth with radiant yellow blossoms. The flowers are fringelike, bearing petals that resemble slender straps, and appear as miniature sunbursts all over this shrub.

A close study of witch hazel would reveal male stamens that open suddenly to release clouds of pollen. The fruits actually take about a year to mature, and open in as dramatic a way: cool nights and dry, warm days cause the capsules to actually curve inward, creating considerable pressure, which makes the fruits eventually explode, sending forth two shiny black seeds as far as forty-five feet. Often, when walking near this shrub in the late fall, you will hear the "ping" of a seed being shot and landing against a leaf or twig with some force. The mechanics involved in witch hazel's dispersal of its seeds are most impressive—and alarming, if the source of this activity is not known!

Equally impressive are the plant's many uses. Iroquois

Indians, from whom many of these uses were learned, made a tea from dried witch hazel leaves, which they sweetened with maple sugar. Early settlers are said to have used its forked twigs as dousing rods to locate hidden springs of water or mines of precious metals.

Witch hazel's reputation as an herbal remedy is known far and wide. The distilled extract from this plant is reputed to be effective in the treatment of internal and external hemorrhages, bruises, hemorrhoids, varicose veins, burns, insect bites, and bags under the eyes. While not a wonder drug or miracle cure, the extract, purchasable even today from most drug stores, is said to be soothingly effective on minor burns, bruises, insect stings, and the like. The high concentration of tannin in the leaves makes them very astringent, and this astringency makes the extract an excellent skin freshener.

Certainly any herbal remedy that has withstood the test of time so well, and still finds a place in many of our medicine closets, deserves our attention to the plant in the wild. [11]

GARDEN WRAP-UP

When summer's over, it doesn't mean the end of the garden yet. Lots of things still are left to do and possibly there is a lot more to harvest than meets the eye.

As the growing season draws to a close, many gardens become weedy jungles. Don't let this happen to yours. Unsightly, weed-ridden gardens harbor disease and insects, and somewhere under all that mess may lie some still salvageable produce.

Get rid of the weeds. Pull them up and burn them. Or if they're not yet loaded with ripening seeds, throw them on the compost pile or till them under. If you're sure the vegetables are all done, you can till them under, too. The added organic matter will benefit your garden soil.

Before you do the whole plot in, take a look at the status of your vegetable plants. There could be quite a few harvestable buds on the broccoli where you cut heads earlier. There still may be some brussels sprouts coming along; and what about those late plantings of beets and carrots you made last July? They might be ready to harvest. Or, you may wish to leave your carrots right in the ground to overwinter and provide a tasty early spring harvest. This, of course, is the way to do it with parsnips, an old Vermont tradition.

Now for the peppers, tomatoes, cucumbers, and zucchini. Gather up what's left, green or ripe, and concoct some "end-of-the-garden" relish. Green beans, celery, cabbage, and even sweet corn can go into this delightful preserve.

Overly large zucchini may be stuffed and baked, or ground into relish, or added to a batch of zucchini bread to be frozen for later use. Those ripe, overly large cucumbers will make great tongue pickles.

If frost hasn't reached you yet, and you have green tomatoes of decent size still on the plants, pick them (vines and all), and hang them up in a protected place where they'll ripen gradually. If you pick the individual green tomatoes, store them in a single layer in shallow boxes or trays in a cool place for slow ripening. Sunlight isn't necessary for them to turn red, and wrapping them in paper is optional and not absolutely necessary. Check them from time to time, discarding those that spoil. With a little luck, you can have tomatoes gradually ripening right up to Thanksgiving time.

Late-planted onions, especially those you grew from seed sown right on the garden, may be coming along now. Those of decent size and those with their tops dying back will be ready to pull. Spread them out or hang them up to dry in the sun for a short while, then store in a dark, cool place.

As you carefully pick your butternut, hubbard, and other winter squash, dip them into a mild chlorine bleach solution to help stave off storage rot. After dipping, air or towel dry and place in a warm spot indoors for a week to hasten curing. After their skins have hardened, you can store them in a moderately cool place for their long winter nap.

With your garden cleaned and tilled or sown to a cover crop before snow flies, you'll be giving yourself a great head start for next year's gardening. [12]

HOME STORAGE OF VEGETABLES

Did you plant a lot of beets and carrots, and then remember that you didn't build a root cellar for storage? Maybe your basement is too hot. Or maybe you garden on a community garden patch and live in an apartment. Maybe you don't have a cellar at all.

Regardless of your problem, don't despair. There are special ways you can carry over your bounty without a lot of extra canning and freezing. You can leave root crops in the ground and mulch heavily with straw or hay so you can harvest throughout the winter, or a portable root cellar with do the job.

First, to preserve root crops for a long period of time, you'll need to provide rather cool conditions (thirty-five to forty degrees F) along with sufficient humidity to prevent spoilage and to preserve texture, flavor, and nutrients. You'll also need a box, barrel, or basket to contain your crops.

A modest-sized container is best. Although round baskets and barrels will do, these are hard to find and difficult to handle. Boxes are best. Look for cartons the size of apple boxes, double thickness and waterproof. An apple box itself is ideal.

Line the box with a loosely fitted plastic trash bag. Now for the packing. Start with a generous layer of sawdust, shavings, vermiculite, or peat moss in the bottom, and add a loose layer of root vegetables. Then alternate layers of packing and roots, topped off with at least two inches of packing medium.

If this appears to be quite dry, sprinkle in a bit of water, then close the top of the plastic bag.

Punch a few holes to let excess moisture escape. It's also wise to open up your "root cellar" package from time to time. Check for proper dampness, and look for any signs of mold or spoilage.

Where to store it? In any unheated room or place where the roots won't be frozen.

Perhaps one box will be enough for you, and different roots can go in together. However, you may wish to store turnips in a separate container. With ample produce, carrots would go into one box, beets in another, and so forth.

All will keep well and provide you with garden produce well into winter. Remember to check your stored vegetables for spoilage on a regular basis. And remember the old axiom that one rotten apple spoils the whole barrel. [13]

LEAF DROP OF EVERGREENS

Do your evergreen plants turn yellow in the fall? If so, you should not be alarmed. This is a natural condition.

Actually, evergreen plants are not really "ever green." They remain green throughout the year only because they do not lose all their foliage at one time as maples and other deciduous plants do. Usually, annual leaf or needle drop of evergreens goes unnoticed because new leaves or needles conceal the old, inside foliage that has turned yellow and brown.

Broad-leaved evergreens such as rhododendrons usually shed old leaves in the spring or early summer after the flush of new growth. Most other evergreens drop their leaves or needles in the fall.

Each evergreen plant has leaves or needles of different ages. The life span of a needle varies depending on the type of tree. White pine, pitch pine, balsam fir, and Eastern hemlock will hold needles for two or three years; red pine for three or four years; red spruce for five or six years; and white spruce and Douglas fir eight years or longer. An exception is the larch or tamarack, which loses all its needles annually, turning yellow each fall and green with its new foliage each spring.

Any factor that decreases the vigor of evergreens, such as a change in the environment, may stimulate premature and excessive leaf drop. Evergreens planted in wet or poorly drained soils will often show an abnormal amount of leaf yellowing on the inside branches. On the other hand, if evergreens are not provided with sufficient water during the dry part of the summer, leaf or needle drop may be earlier and more severe than normal. A lack of nutrition also may cause a short period of growth and early leaf drop.

Most evergreens around the home are grown in sites far removed from their native habitat, and special care is often required at the time of transplanting. Break and loosen the ball of soil surrounding the roots to provide better aeration after transplanting.

Mites frequently lead to nonseasonal needle drop. Generally, the needles of infected trees are off-color, becoming yellowish or brown. A light webbing is associated with heavy infestations. If mites are suspected, hold a sheet of white paper under a branch, and sharply tap the branch. If present, mites will fall onto the paper where they will be more visible.

Some forms of herbicides applied to the lawn or in the vicinity of evergreens may cause sufficient injury to result in needle yellowing and nonseasonal drop. The homeowner should try to prevent this, along with winter injury or drought if possible.

Normal needle drop is a seasonal occurrence, and the symptoms are distributed generally throughout the interior portion of the plant. If you have doubts about accurate diagnosis, examine the leaves and needles carefully. Needles that yellow and drop normally from age may have occasional spots and blemishes. Old needles sometimes show mottled brown coloration from invasion by nondisease-causing fungi. On the other hand, spots or blemishes on the current season's leaves or needles may be caused by insects or disease.

Your local lawn and garden dealer or nurseryman, or your local Extension Service office, can provide help in diagnosing suspected problems, as well as information on the proper care, maintenance, and winter protection of evergreens. [14]

AUTUMN LEAVES

There was frost last night and the squash leaves in Jane's garden are lying in a gray tangle. Luckily she'd already picked the big dark green vegetables that we'll feast on most of the winter. But the maples that march in a row up the edge of our cow pasture are in their glory. There's one old veteran that has become so bright a scarlet it almost hurts to look at it. Farther up is a majestic fountain of deep yellow. A thicket of sumac beneath is doing its own thing by turning itself into a vermilion bonfire. And not to be overlooked are the subtle mauves of the white ashes. I guess there's nowhere else in the world where trees and bushes make such a fantastic tapestry across the hills.

I used to think that color changes were caused by good old Jack Frost, but now I know better: it's all due to hormones. It seems that there's an inner clock mechanism in a tree that says days are getting shorter, winter's coming, and so like householders in the north country, the tree starts to button itself up. It begins by producing a hormone in the leaves, which travels down the leaf stem to the base where it's attached to the twig. There the hormone stiffens the cells making a cork layer. This layer seals off the moisture, keeping it in the tree, for the leaves don't need it now, as with the shortened days, there isn't much photosynthesis, or foodmaking anyway. By cutting across the leaf stem, the corky layer (its scientific name is "abscission layer," from *abscissus,* to cut) stops any more water from coming into the leaves. Without water, the green chlorophyll begins to break down, revealing the yellow and orange pigments that have been present all the time in the leaves but have been overcolored by the green chlorophyll. The corky layer makes it easy for the leaf to fall off when it is no longer needed.

The red of that spectacular maple along our pasture wall is caused by a further step in the process. In these red maples some food or glucose is trapped in the leaves and is synthesized further into a pigment called anthocyanin. Low temperatures and strong light intensities on clear fall days hasten the production of this pigment. You might think that anyone living in these parts a good many years and seeing a lot of falls would grow a little bored with red, yellow, and scarlet, but every year I think the color across our hills is the best yet. [15]

FALL LAWN CARE

Faced with life outdoors all winter in New England, a lawn needs all the help you can give it.

A lawn is not too particular about its requirements, but you can help it survive the winter in good health and emerge vigorous next spring if you follow a few maintenance practices in the fall.

As leaves begin to fall, remove them long before snow arrives. They not only shade the grass during late fall but also become wet and mat down to smother the grass over winter. A push sweeper is the easiest tool, and broom rakes also work well. Some rotary mowers have leaf-mulching attachments. There's no harm in returning pulverized leaves to the lawn unless this aggravates a thatch problem. (Leaves are excellent additions to your compost pile.)

Continue to mow your lawn until growth ceases. Depending on the year, this may be as late as Thanksgiving in warmer parts of the north country. Set your mower as low as possible for the last mowing to remove nearly all top growth. This will avoid the need to rake off dead growth in the spring and will help eliminate snowmold.

If snowmold has been a problem in the past, you may expect repeats on a yearly basis. Even if you don't recognize the disease, you probably have seen its symptoms — patchy dead areas all over the lawn that persist after most of the grass has greened up in early spring. Prevention is possible, but snowmold cannot be cured without remaking the lawn.

Snowmold flourishes on bentgrass, a component of many older lawns. Fungicides applied just before snow arrives will alleviate the problem, but environmental concerns may be enough to encourage you to find solutions instead of treating symptoms.

Shorter days and lower light intensities in late summer and fall, coupled with cooler nights and increased precipitation, bring out new forms of fungus problems, the most obvious of which are toadstools and mushrooms. These are fruiting bodies of soil fungi that flourish beneath your lawn as inconspicuous, white, cobwebby strands called mycelia. Fungi are plants, but as they lack chlorophyll, they cannot make their own food as green plants do. Therefore, they obtain energy from organic sources.

Fungi often grow in arcs or circles around a bit of wood buried in the soil. Their location is evident because they fix nitrogen from the air and share it with lawn grasses, which show greener foliage, called fairy rings. These are especially obvious when mushrooms are showing. Chemical control of mushrooms is usually not necessary. Remove the woody source of food, and the fungi will die.

Fall is also the time of year when powdery mildew becomes apparent. It flourishes in lilacs and many garden plants as well as in grasses. Infections may be severe enough to kill some Merion and several bluegrass varieties, but fescues are quite immune. Lack of sunlight is a key factor in growth of powdery mildew, so look for it in shady places, such as under roof overhangs and in tree shadows. As days shorten in October, you may find it right out in the open, especially if the weather has been cloudy. Fungicides are available at your local lawn center and garden dealer to control this disease, or you can reseed with red fescue and a resistant variety of bluegrass. [16]

WATERING AND INDOOR PLANTS

Overwatering is one of the greatest causes of failure in growing house plants. Most people water as soon as the soil crust dries. House plant roots are usually in the bottom two-thirds of the pot. You should not water until this soil starts to dry out slightly.

To test for dampness in a six-inch pot, poke your finger two inches into the soil. If it feels damp, don't water. Special metering devices that measure soil moisture and indicate when to water are also available at most garden supply stores.

Plants should be watered until the water runs out of the bottom of the pot. This helps wash away excess salts in the soil and guarantees adequate watering. Don't allow the pot to sit in a pool of water, however.

Subirrigation, or watering from the bottom, is a popular and very acceptable watering method, especially for African violets. But at least once a month during the winter and two or three times a month during the summer, water from the top. This flushes out salts and fertilizer residues.

Pots without drainage holes don't have to be watered as often. When you do water, don't overdo it. Remember, there's no place for the overflow to go.

When given a choice, don't buy pots or plants in pots without drainage holes. It's extremely difficult for the average home gardener to grow plants successfully in such containers.

You can use water temperature to regulate plant growth. The principle is simple. Cold water slows growth. Warm water speeds it up. Avoid extreme temperatures, however. Using cold tap water in the winter may stop root and plant growth and damage roots.

Following a regular watering schedule and regulating the temperature and amount of water is the key to keeping your house plants healthy and happy.   [17]

MOSS AND BERRY BOWLS

Want something green and alive to brighten cold, gray, winter days? If potted plants take up too much time or space, you might try a berry bowl or one with native ferns and mosses.

Such bowls are not new. People have had tiny gardens in glass jars and bottles for ages. What's new is the variety of containers for your plants and the many tiny wild plants to put inside.

For an old-fashioned berry bowl, you simply need a small fish bowl or a large-mouthed, squat glass jar of some sort. Pickle, peanut butter, candy, or cookie jars are useful. Instant coffee jars are great.

Next, you'll need plants. If you can get outside, look for partridgeberry (*Mitchella repens*) mosses, and tiny ferns. The easiest kinds of moss to work with lie flat on logs and rocks in moist, shady woods. Peel off small patches of these mosses, but take only enough for your immediate needs.

After washing the bowl, line the lower, inner half with a hollow nest of green moss. Be sure the green side faces out through the glass. Fit and shape the moss to fit with scissors before inserting. A table fork, large tongs, or tweezers are helpful in positioning this moss liner.

Next, pour into the center of your moss cup a spoonful of activated aquarium charcoal, a few spoons of gravel, and enough sterile potting soil or African violet soil to bring the contents almost to the top edge of the moss.

Now carefully insert sprigs of partridgeberry. With any luck you'll have found some with vivid, scarlet berries to add an excellent color accent to your bowl. Even if the sprigs of partridgeberry are not rooted, they'll soon develop rootlets in their new environment.

Now add bits of moss among the berry sprigs to cover the surface of the soil, then a tablespoon or two of water, just enough to dampen the soil. Cover the bowl with its original lid or plastic wrap.

Your covered bowl is now a self-contained garden-in-glass, where the air and water are continually recycled. Your moss ferns and berry vines will continually renew the air inside the bowl. Moisture given off by the living plants will condense on the inside of the glass and trickle down to be taken up again by the thirsty roots. This sort of system operates successfully inside any terrarium.

Sometimes the balance isn't quite perfect and you'll need to make adjustments. If too much moisture condenses on the inside or if the foliage looks damp and decayed, open the bowl and air it out. No mist on the glass, shrivelling, brittle foliage, or brown, crisp moss say "a little more water please."

A moss or berry bowl could get you started on gardening in glass containers. The bowls make interesting gifts, and there are many good books on the subject. [18]

Mammals

ANIMALS

It is not without significance that the Bible describes the genesis of the birds, sea creatures, and animals as taking place before the creation of mankind. The story certainly can be read as describing the preparation of the world for Adam, who then establishes his dominion by giving the other creatures names. To name something, or to be able to call it by a name, often represents a claim to power over it. In many cultures names have been powerful magic spells.

The animal kingdom is perhaps even more central to a conception of the world order that is older than that of the Bible. Before the time came when humans turned them into pets and beasts of burden or funneled them into controlled preserves and labeled them "endangered species," the animals were the true rulers of the world. When the puny population of the human race could have been trampled out of existence under the advance of a single great herd of bison or reindeer, it was clear indeed who were the masters. Mankind knew well enough that the animals were stronger, fiercer, cleverer, and certainly more beautiful than themselves.

It is with eyes attuned as much as possible to this view that we should look at the animals as they are shown by the artists of the primitive world. Many of them show the animals as ancestors of man.

The jaguar, one of the great feline predators, prowls through all the ancient art of the Americans, a fanged and clawed image of terror.

Indeed, it is a measure of their power that animals often have been seen as man's true ancestors, our fathers and mothers of long ago. Consequently, hunting was often forbidden to the animals' putative descendants; in some cultures, men were obliged to ask the animal's permission for the hunt in advance and to apologize to it for its loss of life afterward.

Masks in animal form often express this kinship; the man becomes his ancestral beast and reenacts its deeds. Sometimes he impersonates animals or animal-like heroes who taught humanity essential skills. Very often in Africa the mask represents not so much the animal itself as qualities associated with it that are in turn associated with gods: the wildness of the gorilla, the power of the buffalo. Emblematic features are combined to represent the totality of a god's attributes. They also represent the human being's haunting discontent with his own powers, his longing to make them ever greater. [1]

GROWING OLD

A field mouse in the wild is lucky to live to be a year old. Yet, its cousin, the pocket mouse, may live to five years of age. Why do some animals live longer than others?

According to *International Wildlife* magazine, scientists are studying senescence, or aging, in wildlife to find out the answers to these questions, and in turn, to better understand the processes involved in human aging.

The answer to the mouse mystery is simple. Some small mammals may live longer than others of similar size because they spend part of their lives in hibernation or in dormancy, their bodies barely functioning. Thus the pocket mouse, which spends the hot desert season in a state of torpor, will live much longer than the on-the-go field mouse.

Until recently, almost everything we know about animals that live to be old came from zoos. Longevity records for zoo animals include: a seventy-seven-year-old elephant, a sixty-five-year-old vulture, a forty-nine-year-old hippopotamus, and a thirty-eight-year-old zebra. Cold-blooded animals last even longer, with tortoises reaching one hundred and fifty years; sturgeon one hundred; carp fifty; and toads thirty-six.

Unfortunately, zoo records can be misleading, for animals tend to live longer in the security of a zoo than they do in the wild. Still, such records do give us an appreciation for the potential life spans of some species.

Animals that are most likely to live to be old in the wild are those that, as adults, have no predators except man. Most of these are large—elephants, hippos, rhinos, lions, tigers, and wolves. But some, such as the Galapagos tortoise, survive not only because of their large size, but also because of their isolated habitat.

Scientists have found that a wild animal's age can be estimated in many ways, such as by inspecting the lens from a rabbit's eye, the thickness of baleen plates in the mouths of whales, the annual rings in the horns of sheep and antelope, and tooth wear, or, in some species, rings in a tooth. Teeth are, in fact, one of the vital considerations for life in the wild; research shows that most wild animals live only as long as their teeth hold out. [2]

YAWNING ANIMALS

Anyone who has been embarrassed by a wide-mouthed yawn in the middle of an economics lecture, a budget discussion, or a friend's emotional outburst knows that the yawn clearly delivers one of two messages: boredom or fatigue.

It's not so simple in the animal kingdom. Animals do yawn—but for a whole variety of reasons. The animal's yawn may signal courtship, aggression, warning, or a need to create order within a group.

The conventional yawn occurs when breathing slows down because of fatigue, inactivity, or lack of sleep. Opening the mouth wide creates a sudden intake of air that sends oxygenated blood to the heart. The heart, in turn, rushes blood to lethargic muscles, reviving a sluggish system.

Some animals do yawn for conventional—or physiological—reasons. For instance, the javelina (wild boar) yawns to rejuvenate himself whenever he leaves the shallow depressions in the desert where he sleeps.

But for other animals, yawns are mere weapons in territorial battles. Lizards and fish commonly threaten unwelcome invaders by opening their mouths. Bears and wolves bare their teeth before an attack. The hippopotamus displays perhaps the most spectacular aggressive yawn. Two hippos trying to settle a dispute stand head-to-head and flash two-foot-long teeth in giant yawns. They have been known to match yawn for yawn over several hours before settling a quarrel. Baboons also issue threats with yawns— often to establish pecking orders within groups.

Lions, on the other hand, spread their jaws to calm tensions within a group, according to Randall Eaton, a professor of animal behavior at Western Wyoming College. Female lions yawn frequently, almost in unison, to distract potential aggressors from their newborns.

Some animals even yawn to invite clean-up crews into their mouths. A case in point is the crocodile, who can't move his tongue well enough to clean his teeth. He instead spreads his jaws wide so that certain birds can dine on leftover food among his teeth.

When animals get the irrepressible urge to yawn, they seem to know, as man does, when to stifle it. Lions, baboons, and other animals that live in groups have perfected the technique. When they don't want a reflex yawn to be perceived as a threatening gesture, these animals look away or cover their teeth with their lips. Lower-ranking animals turn away from higher-ranked animals when they can't suppress a yawn. [3]

ANIMAL LANGUAGE

No one has ever doubted that animals have wonderfully diverse and ingenious communications systems, which they use to advertise such basic needs as danger, hunger, and readiness to mate. Some of these systems are by no means simple. The female of one species of firefly, for example, has learned to imitate the flashing signal of another species to lure in the alien males, eating them as they arrive. But such abilities, no matter how ingenious and complicated, hardly qualify as language.

Some animals, however, seem to exhibit humanlike inventiveness in the way they communicate information. Ground squirrels, for instance, employ two classes of alarm calls to indicate type of predator. Paul Sherman of Cornell University reports that when these social rodents hear the alarm call that signals the approach of a bird of prey they dive into the nearest cover. But the call that signals a digging predator, such as a badger, causes the squirrels to bypass perfectly good nearby burrows as they head for one with a backdoor escape route.

Scientists have long known that the humble honeybee defies another stricture—that only humans can refer to things or events distant in space or time. The honeybee uses its spectacular waggle dance to announce to other bees the location of far-off food, water, or nest sites it visited hours or days earlier. Another sobering similarity to human language is that the bees have dialects: a waggle that means fifty meters to an Austrian bee means ten meters to an Egyptian one.

Linguists counter that what bees do is programmed entirely in their genes and that, to qualify as language, a communication system must be transmitted culturally. And in some other nonhuman creatures it is. In 1964, for example, Peter Marler of the Rockefeller University discovered that white-crowned sparrows in California sing in regional dialects. The idea that physiologically identical birds could have songs that vary among cultural groups intrigued Marler, and he began to map dialects all around the San Francisco Bay area. He found that not only are there large differences in the songs of birds from Berkeley, Inverness, and Watsonville, but there are subtle differences on opposite sides of each town. [4]

INSULATION

Fat can influence body heat. A layer of blubber allows a pig to maintain its thirty-eight degree C internal temperatures even in cold weather, and encases sea mammals like whales, walruses, and hair seals so they can survive in icy seas. There is even a fat especially adapted for producing heat. It is called brown fat, or adipose tissue, and is present in many newborn mammals, including humans. These brown fat cells generate much more energy than white fat cells, through oxidation, and there is abundant blood flow through them to circulate the newly produced heat.

But why doesn't the skin of these blubbered animals, live tissue after all, lose all its heat to the air and the ocean?

And why, for instance, don't the noses of the caribou, or the feet of Eskimo dogs release a potentially dangerous amount of body heat? An amazing system prevents this heat loss; it is an exchange of heat from the warm outgoing arterial blood to the blood on its way back to the heart from the outer extremities. Thus the blood reaching the cold skin or nose or feet is already cool and doesn't impart heat to the exterior. Also, the blood that is warmed as it returns to the heart does not cool the vital inner body organs. This miraculous exchange of heat occurs in the *rete mirabile* (wonderful net), a network of small arteries and veins near the junction between the trunk of the animal and its extremities.

There is another way to keep the heat that the body has worked so hard to generate. In a way similar to closing off less important rooms in a house, the body can shut off the blood supply to the outer extremities. This is called vaso-constriction and may explain why your hands and feet get cold. The body's metabolic mission is to keep its vital organs at a constant thirty-seven degrees C or ninety-eight point six degrees F. When extreme cold threatens internal chilling, the body automatically constricts the arteries leading to distant parts of the body in order to conserve heat and energy. [5]

MIGRATION

The African elephant is the largest living land mammal in the world. These great creatures make regular seasonal journeys of up to several hundred miles. Their movements are related to climatic conditions, food and water supplies, and the breeding cycle. In some areas, the elephants' journeys have been so regular that their massive frames pounding over the same ground every year have created definite highways that are almost bare of vegetation.

Breeding parties of old bulls and females have been observed on long migrations. In 1922, the zoologist C.W. Hobley wrote an account of such a breeding party in southern Kenya. He estimated that the elephants took three years to complete a round trip of four hundred miles. After courtship and breeding the young were born in the Marsabit Forest, the favorite breeding ground of elephants in the area.

In North America, as in Africa, the land mammals that migrate are the hoofed mammals. Many of their movements have been prevented over the years because man has fenced off the prairies, introduced livestock, and built roads and railways.

The most spectacular animal migrations in the New World are those of the caribou. At the beginning of the twentieth century, about one and three-quarter million of these large deer roamed the wilderness of northern Canada. But excessive hunting, forest fires, and competition for their food from reindeer — a variety of the caribou introduced from Europe — took their toll. By 1955, the caribou population had dropped to 278,900, but through an intensive program of conservation the numbers have increased, so that there are now over 375,000 wild caribou in Canada.

Many caribou breed during spring on the tundra, or barren lands, of arctic Canada. These animals belong to the subspecies known as the barren ground caribou. The Canadian tundra, which occupies an area almost twice that of France, provides enough food for the caribou during the brief summer season, but when the arctic winter sets in, the mosses and lichens die off and the ground becomes covered with an ironhard layer of ice. The caribou herds are then forced southward in search of food.

From July onward, the caribou begin to move south toward the timberline — the northern limit of trees. During the harsh Canadian winter, they find shelter and food within the great expanses of the northern forest.

By March and April, the caribou are once again on the move. This time they are traveling northward, returning to their breeding grounds. Their annual journeys average seven hundred miles. [6]

THE HUNTER AND THE HUNTED

Every animal on earth shares the same problem. It must get enough nourishment to keep its body machinery going or else face death. The source of all the nourishment is the sun, whose energy is trapped by plants in photosynthesis. Herbivores are animals that depend solely on plants for their food. Carnivores depend on herbivores to supply their food. Omnivores are flexible animals, in that they eat both plant and animal matter. The problem of getting enough food to eat is quite different for an herbivore and a carnivore. An herbivore's food source is stationary and often fairly abundant. The problem is how to get enough of it to support the animal's body weight rather than how to catch it. Carnivores have to work harder for their food. Meat is a more concentrated food source but much harder to catch. Predators must develop many techniques to insure a stable food supply.

The interaction between predator and prey, the hunter and the hunted, is a basic, life-or-death situation in nature. Neither the hunter nor the hunted are right or wrong, evil or good. They just *are*. Both have evolved fascinating adaptations and habits to help them be more successful at surviving.

Predators are often looked upon as nasty fellows with blood dripping from their fangs. However, predation is an absolute necessity in the ecological community and is as important to the prey as to the predator. Hunters invariably go for the easiest prey they can find. It is of no value to them to exhaust themselves needlessly on a prey that is too difficult to catch. So they usually weed out the fringe members of a species—the old, diseased, or very young. If these animals were not eliminated, they would decrease the chances of the rest of their group's survival by competing for food and space. Overcrowded conditions create animals that are more vulnerable to disease, parasites, and psychological problems. Abnormal behavior, such as unnatural aggressiveness, may develop, which jeopardizes the mating instincts and the care the young receive. When the weaker members are removed, it is easier for the remaining animals to survive and reproduce. The hunters are unconsciously applying a biological brake to help keep the animal populations near an optimum level. Both the hunter and the hunted benefit. [7]

BAG BALM

There's a product on the shelves of American feedstores that cows have been mooing about for years. It's called Bag Balm, and it's meant to help heal chapped udders. But cows, it turns out, aren't the only ones impressed by Bag Balm's healing power. People are rubbing it on everything from their hands and feet to their bedsprings and machine guns.

"I really can't say why people use it," says sixty-nine-year-old John L. Norris, Jr., whose company in Lyndonville, Vermont, has made Bag Balm for seventy-five years. "As far as I'm concerned, it's for veterinary use only, just like it says on the can."

Sixty years' worth of Bag Balm testimonials are piled on a desk outside Norris's office. Marathon runners report that Bag Balm prevents blisters on the big toe and outer borders of the feet. Skiers, particularly in the New England area, coat their faces and hands with the thick yellow salve. Kayakers, whose skin is exposed to sun, wind, and salt water, consider Bag Balm standard equipment.

Actually, Bag Balm is a combination of four ingredients, one of which is pine oil, giving the ointment an aroma rather like sweet turpentine. Lanolin and petroleum jelly, which can be found in many over-the-counter lotions, creams, hair conditioners, and other cosmetics, are moisturizing agents. Whether in cows or people, chapped and cracking skin is a result of too much water being lost from the skin's outer layer. For people the cause is often overexposure to sunlight, dry winter air, or harsh cleansers. And cows frequently suffer because their udders are subjected to soapy water, disinfectant, and up to three milkings a day. Bag Balm's two emollients lubricate the chapped areas, seal in moisture, and protect against further exposure.

Bag Balm enthusiam isn't limited to New England. In California, a Bag Balm T–shirt was designed for a recent marathon. And in Marysville, Washington, the Cedar Crest Golf Course hosts the annual Bag Balm Invitational in autumn.

While all this hoopla would be welcomed in most businesses, Norris appears uncomfortable. He claims he'd like to keep the company small, continuing to sell Bag Balm strictly for bovine use. "You have to realize that we're talking about plain country medicine," says Norris. "Bag Balm takes care of cows. If anything else comes our way, we just say thank you." [8]

THE BEAVER

Other than man, the beaver itself may be his worst enemy. Without keeping their numbers in check with an annual harvest, the colony soon eats itself out of house and home. Although the beaver is probably the least prolific rodent, breeding only once a year and having an average litter of four kits, the colony expands rapidly. A colony consists of three generations—the adults, the year-old juveniles and the current year's kits. This means there are at least eight, and more probably ten or eleven, members in an established colony. A lot of foraging, tooth-sharpening and tree-cutting must occur to sustain it.

An unchecked colony soon finds that the trees and vegetation it needs for survival are being consumed faster than they are being replaced.

When this happens younger members of the colony are forced to migrate to a fresh sourch of food. But these migrations can be fatal. A beaver in the water is a match for any predator. But on land it is clumsy and vulnerable.

A normal size adult beaver weighs fifty to sixty-five pounds, with an overall length of forty inches.

The hind feet are much larger than the front. The thin, duck-like web of skin between the toes greatly increases the beaver's mobility in water. The second toe on the inside of each hind foot has a split toenail. These nails are used to groom and waterproof its coat with an oil extracted from its vent.

The tail of the beaver is also unique. It is flat and covered with scales. The tail functions as a counterbalance and a prop for the beaver when sitting to fell a tree or groom itself. It also serves as a rudder when the beaver is swimming.

As with all rodents, teeth of beaver grow continuously. Chewing not only prunes the teeth but keeps them sharp.

The eyes of beaver are small and protected by transparent lids that cover the eyes when the beaver is submerged. This allows the beaver to see underwater without getting any irritating substances in its eyes. The ears and nose of the beaver can be closed tightly when it dives.

Beaver possess two types of body glands. One secretes an oil they use to waterproof their fur. The other secretes sweet-smelling castoreum, used in two situations. One is to mark its territory to notify noncolony beaver they are entering claimed waters, and the other is to use as a calling card to find a mate. [9]

RED SQUIRREL

Venture into an area occupied by the red squirrel and the little animal will most likely bark, spit, sputter, and growl at you from among the branches, observing your every move with a suspicious eye. And should you show designs on his storehouse, attempt to look into his nest, or even come a little too close, he will fly into a rage and with convulsive movements stamp his feet and bounce about scolding you with all the fury he can muster.

The red squirrel is a house builder of no mean talents. His favorite home is a nest in a hollow tree, an impregnable little fort and a dry, weatherproof house. If he is unable to find a suitable cavity, he will construct a roundish nest of leaves, pine needles, shreds of cedar or other bark, moss, dry grasses, and twigs in a whorl of many branches or in a witch's broom usually thirty or more feet above the ground (but sometimes as low as six or eight feet). It is ingeniously contrived and surprisingly wind- and rainproof.

He also constructs an underground den, usually beneath a tree stump so it can provide him with a strong roof.

The home range of the red squirrel is limited to some five hundred to seven hundred feet across, beyond which he rarely ventures except perhaps at mating time. Adapted in many ways for an arboreal existence, he spends most of his time in the trees running up and down the trunks, and along the branches with surprising speed and endurance, frequently zipping through the air from one tree to another with leaps of six to eight feet.

Most of his food he obtains from succulent growing twigs and buds, flower parts, and seeds from shrubs and trees, the latter providing the greater portion. But also included in his diet are insects of various kinds: the pupae of moths, hornets, wasps, and bees, the larvae of bark and wood-boring beetles, plant lice, and grasshoppers, not to mention young birds and eggs occasionally. [10]

ANTLERS

To all appearances the antlers that adorn the heads of male members of the deer tribe seem to be permanent as the animals themselves, but they are in fact only temporary appendages. Each year they are shed and replaced with a new set.

Beginning in the spring as soft, swollen pads on the skull, they soon lengthen into clublike structures. While growing they are covered with a fuzzy skin called velvet, beneath which blood circulates through a network of vessels. The tips are bulbous, the entire antler tender and easily damaged.

In two months they begin to show the general shape of the antlers to come—in some instances mere spikes, in others elaborately branched or palmated affairs. Four to six weeks later they reach full size.

At full size, they undergo a surprising transformation. Beneath their furry covering the antlers harden and the blood supply stops. The velvet, now dead and dry, peels off in strips, aided by the buck's vigorous rubbing against trees and bushes. The antlers are now bone-hard, with furrowed bases and pointed tines.

For a few weeks in autumn they resist the incredible punishment of head-on rutting clashes. Then one day, when the mating and fighting urge has passed, they suddenly drop from the buck's head, leaving only a pair of bony bases from which next year's set will grow.

The cycle is the same with deer, moose, elk, and caribou—all of the deer family. Although the moose's huge antlers may span seventy-six inches and weigh seventy pounds, and the elk's majestic ones measure six feet or more in length, they nevertheless attain these impressive proportions in the short period of three or four months, making them the fastest growing animal tissue known.

What happens to the shed antlers? Within a few

months they have usually been reduced to unrecognizable fragments by decay and the incisors of mice, squirrels, and porcupines, to whom they are a welcome source of salt and calcium. [11]

THE RED DEER

Name a domestic animal raised on two thousand farms in the Southern Hemisphere whose meat and soft hair are in great demand. Sheep? Llamas?

Neither. The animals are red deer—the world's newest domestic stock—and raising them is a booming business in New Zealand. The red deer, originally captured from high-country forests, are joining sheep in New Zealand pastures.

All domestic animals are descendants of wild species. But what's different about New Zealand's red deer is that the domestication has taken place in a few years, rather than millenia. The first red deer farm was started in 1970. Today, the country has a herd of more than 180,000 animals.

That growth is even more remarkable when one considers that the first deer arrived from Europe in 1851. Previously, plants and flightless birds had reigned unchallenged in New Zealand, and people had been the first mammals to arrive in about A.D. 950.

In those New Zealand surroundings, the deer found abundant food year-round—and no competitors. Within a year, their breeding had adjusted to suit the new seasons of the Southern Hemisphere; they survived winter easily, grew quickly, and bred early in life. As its population increased, however, the deer's overgrazing denuded many slopes. After deer trampled out vegetation, soil slid into valleys, causing floods and clogging hydroelectric dam reservoirs with silt.

What seemed like an ecological disaster was to become an economic boom as farmers realized the profit potential in domesticating deer. Since the first deer farm was started, New Zealand's animal scientists at the Invermay Animal Research Station have uncovered some interesting facts:

—Deer produce over 40 percent more meat per acre than sheep or cattle.

—Deer raised in captivity may achieve commercial body weight within fifteen months, compared with thirty months in the wild.

—Pasture that supports one dairy cow will support four deer.

New Zealand deer farmers hope that the venison industry will one day rival the country's massive wool trade. Venison is a highly prized meat, considered a luxury item in many restaurants.

Venison is not the only product of New Zealand's deer farms. "Velvet," the soft hair that covers a stag's antlers and is shed every year, is another. One farmer claims that velvet from one stag is worth as much as the wool from twenty four sheep. In China, Korea, and the Soviet Union, velvet is used in treatment of heart disorders and anemia. Doctors there say it dilates blood vessels and stimulates production of red blood cells.

Ken Miers, the director of the environmental division of the New Zealand Forest Service, says the deer are now "absolutely under control. It's the best thing that ever happened, as far as we're concerned." [12]

HIMALAYAN ANIMALS

It's not deer and antelope that roam high atop the Himalayas, the snowcapped mountains wedged between India and Tibet. Instead these mountains, which rise to altitudes of 20,000 feet, are homes to some of the rarest animals known to man—the tahr, argali, serow, and yak, among others.

Scientists only recently have begun to understand these high living animals. They're learning now how some eleven species of ungulates, or hoofed mammals, have adapted to their harsh, inaccessible, and arid habitats, where temperatures rise to one hundred degrees Fahrenheit and plunge to minus forty-five.

The area is so remote that some of these creatures were not known to science until the nineteenth century. Each species has its own niche, usually determined by climate and vegetation that vary with different elevations.

The Himalayan tahr's ability to clear a six-foot obstacle from a standing position makes it the champion high jumper of the highland herds. He has also developed climbing skills to maneuver amidst the rocky crags of his mountain home. A close relative of the goat, the tahr is found between elevations of 7,000 and 14,000 feet.

The argali's spectacular horns, which can reach a length of seventy-five inches make it one of the world's most prized big game trophies. Hunters who go after these animals must be in top physical condition because the argali doesn't descend below 15,000 feet.

The serow has been described as a cross among a cow, donkey, pig, and goat. These stocky animals, which stand about thirty-two inches high and weigh two hundred to three hundred pounds, also have the reputation of being quite aggressive when brought to bay. The serow prefers precipitous, wooded gorges at elevations between 5,000 and 12,000 feet.

The king of the Himalayan mountains is the yak, or "grunting ox," by far the most ungainly looking animal of the Tibetan highlands but also the hardiest. It claims the highest peaks for its home, from 14,000 to 20,000 feet. Domesticated for centuries, yaks are both the cows and the horses of the highlands. They provide not only transportation but also milk and meat. Their droppings are an important source of fuel. No other animals can surpass the yak's ability to cope with deep snow or swim icy waters while carrying a heavy load. [13]

THE BUFFALO GRAZE, AGAIN

That the buffalo somehow managed to stagger out of the last century into our own is a bit of a miracle; or, if not, then it was nothing less than colossal good luck. No one will ever know how many of them were out there when the first paleskin scouts poked west toward the shining mountains for beaver. The prevailing guess is sixty million. And there may have been more than that in pre-Columbian times, before hunters acquired the horse, and before the buffalo's range drew inward to the Plains from Oregon's Blue Mountains and the Appalachian highlands in the East.

About ten years, roughly the 1870s — less than a human generation — is all it took to tumble the millions into hundreds, thanks to the railroads, the long-range Sharps rifles, and an inordinate national appetite for robes and hides. By 1903, the zoologist William T. Hornaday could feel some confidence about the accuracy of numbers; that year he counted 969 buffalo remaining in the United States. With somewhat less confidence in their latest figure, census takers nowadays put the purebred North American buffalo population at 80,000. By all counts, the numbers have nowhere to go but up. [14]

MULES

A mule will never be able to reproduce itself. A male or "horse" mule is born sterile, although he has genitals and must be gelded, for he has the instincts and drive of a stallion. Similarly, a mare or "Molly" mule comes in heat, but if bread to a stallion or jack either will not conceive or will abort.

Hinnies, which are the offspring of stallions and female donkeys called "jennets" or "jennies," also do not bear offspring. These creatures are fairly rare and are smaller than mules. The overly simplified explanation why mules and hinnies cannot reproduce themselves is that they lack the proper number of chromosomes, those bodies within sex cells that must match up, perfectly, two by two, before new life can be created. An ass or donkey has sixty-two chromosomes or thirty-one pairs; a horse has sixty-four, or thirty-two pairs; while a mule or hinny is the odd-animal-out with sixty-three chromosomes.

A mule can be stubborn, and a mule can kick your brains out, but these propensities have been exaggerated beyond reason, say mule men. A mule will kick when provoked or frightened, but so will a horse. Horses can be stubborn, too.

We speak of "horse sense" when we should, more properly, speak of "mule sense," for common sense is the mule's greatest asset. If, for example, you turn a hungry mule loose near an open corncrib, it will eat until it is satisfied and then stop. Neither will it founder from overdrinking. Under the same circumstances, a horse is likely to go on eating until it becomes ill. If you overload a pack mule, it will buck the load off or refuse to move. A mule knows what it can carry and what it can't carry. If a mule becomes tangled, say, in barbed wire, it will wait patiently for someone to come along and extricate it. A horse in the same fix is likely to tear its hide apart in a panicky effort to get free. A mule can stand the heat better than a horse, is stronger pound for pound than a horse, and is heir to far fewer ailments than a horse. A mule is what farmers call an easy keeper, an animal that is cheap to maintain. [15]

MEET WILDLIFE ENEMY NO. 2

Homeless dogs and cats are predators. Some dogs and cats manage to survive in garbage dumps. Some can live on the edge of a city or town by raiding garbage pails. And automobiles, along the nation's 3.7 million miles of roads, kill about one million wild animals of all kinds every day, and this is another source of food for scavengers.

Dogs and cats kill when they must, and sometimes even when they don't have to.

First, they kill wildlife. For example, many farmers own hounds that are seldom confined, and with the hound instinct for the chase, these dogs are particularly harmful to deer, often trailing them just for sport. Feral cats and dogs play havoc with ground-nesting birds, young and inexperienced birds, and small mammals.

On several occasions, in Montauk, New York, a pack of dogs has been seen to run deer into the ocean surf. They hold their death watch on the winter beach until incoming tide and freezing temperatures force the terrified animals ashore and into their jaws.

There is a second way in which feral dogs and cats injure our wildlife; although it is indirect, it is devastating. Feral animals turn on livestock. They are often a major factor in the continuation of predator control programs.

There is not a single dog in Iceland's capital city of Reykjavik — a city of 90,000 people — it is now against the law to own a dog in Iceland unless you need one to herd livestock. And Iceland is the oldest democracy in the world: its Parliament first sat in A.D. 970. If it can come to pass in one democracy, it can in another. Pet owners take heed.

We own cats and dogs because we love animals — most of us, at any rate. It would be ironic if that love plus just a small measure of carelessness were to provide the straw that breaks wildlife's back. A feral dog and cat population very much larger than today's might provide that straw. It could become the turning point for some species in the years ahead. [16]

COYOTES

There seems to be considerable discussion and some confusion about the origin and status of coyotes in the Northeast. Are they merely feral dogs, western coyotes, displaced wolf cousins, or coy-dogs?

The Eastern coyote (*Canis latrans var.*) is a distinct and thriving species in its own right, which has helped to fill a predator niche left vacant by the demise of the wolf. Actually, coyotes are more omnivorous than strictly carnivorous, and much of what they get is scavenged. They help to clean up the numerous deer carcasses that are casualties of harsh winters. Their diet is really quite foxlike, consisting of small rodents, plant materials, rabbits, and insects. A coyote shot in Sudbury, Vermont, for strangely gnashing its teeth and jumping about, was found to have a stomach filled with red-legged grasshoppers.

At the turn of the century, coyotes were unknown east of Wisconsin. It is believed that an Eastern coyote is related to the northern subspecies of the coyote (*Canus latrans thamnos*) found in northern and midwestern states and south central Canada. As the coyotes moved east and picked up some wolf blood, our Eastern coyote resulted. (The wolf blood accounts for its large size.) Vermont's first coyote appeared in 1948, and the species has increased significantly during the sixties and seventies.

What's a coy-dog? These coyote–dog hybrids do occur, but are not thought to really affect the coyote population, because coy-dog females have their pups in midwinter, which is a poor time for survival and certainly not advantageous to the species. Dog genes may contribute to a single generation but wild-to-wild breedings are needed to add to the coyote bloodline.

The coyote is shy, adaptable, helpful, and no threat to man. With time, its necessary predatory role is even becoming appreciated. [17]

THE BIG, GOOD WOLF

Like man, the wolf is a predator, an animal that must kill for food. Predators generally do not kill other predators for food; and wolves as a rule do not think of people as meals. In the wild, wolves kill large, plant-eating mammals such as deer, moose, elk, caribou, buffalo, mountain sheep, beaver, and some wild goats.

Wolves have learned to hunt by varied methods. The pack may use a relay system, in which an animal is singled out and each wolf takes a turn chasing and wearing down the prey. The target is usually old, sick, weak, or very young. The wolves will chase this physically inferior animal until it can be brought down and eaten. In snowy weather, wolves look for prey trapped in drifts. Of course, the easiest catch occurs when the pack discovers a carcass. No part of the prey goes to waste, as the wolves return time and again to clean up remaining meat.

When wolves kill, the entire pack eventually takes part in eating the food. Pack leaders usually eat first, though, consuming the delicacies (intestines, liver, heart, and other internal organs). Scientists have found that wolves are successful in making only seven kills in a hundred tries. Wolves have been known to go seventeen days without food in the wild. Scientists also have discovered that wild wolves can eat up to twenty pounds of food at one sitting, and that approximately ninety percent of the wolf's diet consists of meat.

For a hunter like the wolf, each part of its body serves an important purpose. Long legs enable wolves to run at a top speed of forty miles per hour. In the search for food, wolves often trot at a five-mile-an-hour pace that can be kept up for fifty miles, or half a day. Sturdy leg muscles allow them to leap, slashing at their prey again and again until the kill is made. Wolves have been observed jumping eight to ten feet straight into the air from a sitting position. Wolves have forty-two teeth, and can exert jaw pressure of a thousand pounds per square inch. An adult wolf can crush the thighbone of an elk, moose, caribou, or deer. The wolf's paws are slightly webbed and splayed to balance its body weight on ice and snow. On ice, wolves use their long nails to avoid slipping. In water, the paws spread for paddling.

Wolves band together in a family structure called a pack. But the pack is not only their family; it is also an organized hunting unit. The leader of the pack (called the "alpha") is usually a male, and within the group each member knows its place in the hierarchy. The pack leader does not always rule alone; his mate (or the alpha female) takes part in leading the pack.

Age or personality determines who becomes a pack leader. Older wolves are often better equipped to lead, as they may be better versed in using body language and sounds to dominate other pack members.

When age does not determine the leader, personality becomes important. Wolves are as individual as people, each displaying different character traits. The wolf with the dominant personality will exert his authority over all others, and become the leader. But this does not happen overnight. While they are young pups, wolves fight mock battles with one another. As they grow older, one pup backs down less frequently; he will some day display those physical signs that make him dominant.

Many pack leaders are neither the largest nor the strongest animal. Being a leader is more mental than physical, and the smallest wolf can become boss.

Just as there is a pack leader, there is one wolf (called the "omega") at the bottom of the hierarchy. When approaching other wolves, the omega eagerly licks and nuzzles their faces as a sign of respect and affection. This submissive wolf makes its body appear smaller by standing in a crouched position with its head cast down or to the side. Its ears lie flat against its skull, its tail curls low or tucks between its legs.

Wolves also communicate by howl, whimper, growl, and sometimes bark. One type of howl calls the pack together: a deep howl, followed by a few barks. Another sort tells of a wolf suddenly separated from its pack: beautiful, mournful, it begins on a high note and quickly drops in pitch.

Wolves and dogs are both members of the same family, but wolves rarely bark. When they do, it is a warning or challenge. The alarm bark is short, unfriendly, maybe repeated once or twice to let the pack know something is wrong. The challenge bark often follows an alarm: a series of drawn-out barks. It tells trespassing wolves or strangers to stay away, that they are not welcome; it can deter intruders at the boundaries of the pack's territory. [18]

A FLASH OF MINK

Even those who knew where the mink lived seldom saw her.

A quick flash of brown, almost black in the dimming light of evening, was the most that one could expect when watching for the mink. Like most mammals, she reversed the human schedule, preferring night to day for trips outside the home.

The mink was known as "she" because male mink lack the tenacious commitment to one location that this mink held. Female mink enjoy a home with a river for a front lawn and perhaps a sandbar for a backyard. Male mink are travelers that have a district that may cover several miles. Males have been known to travel fifteen miles in a night.

If one were to guess, one would say that there are more mink today in New England than there were fifty years ago. The mink population, however, probably is hard-pressed. The animals give much attention to their coats, therefore they do not do well in polluted water. The invisible chemicals that invade rivers also work against the mink's future. The chemical soup known as PCBs exists in parts of southern New England, where it causes not only stillborn kits but also death among adults. PCBs contaminate fish that mink eat.

Mink will have their young in late April or, more commonly, in the first week of May. The kits are about the size of a human little finger, and there are usually three to four of them. The young are weaned when about six weeks old but remain with their mother until late August when the family breaks up. Mink are antisocial. Except when the young are being reared, the rule is one mink to one burrow.

Young mink start out life in a white coat. By the time it leaves the burrow, the average young mink will be wearing brownish fur. As members of the weasel tribe, mink belong to a group of animals capable of responding to seasonal changes by switching from the brown of summer to the white of winter. Mink themselves, though, do not experience seasonal color changes. The ability of their cousins to do so may account for the variability of mink color forms in captivity. [19]

THE WEASEL

As a predator, the short-tailed weasel deserves our respect, in spite of the bad publicity it has received over the years. This swift carnivore has inherited the reputation of being wasteful, destructive, and dangerous, to say nothing of bloodthirsty. In fact, they, like all predators, are simply trying to survive, and in so doing contribute significantly and effectively to the balance of nature.

A weasel will take the prey that is easiest to kill—the smallest, slowest, weakest prey available. This not only allows the weasel to survive, but also keeps the prey population healthy and strong.

Small rodents represent the mainstay of a weasel's diet. Chickens, rabbits, and other comparatively large creatures are the exception. Having killed its prey, it is true that a weasel will eat only what he needs at the time, not being able to store much fat. Its bloodthirsty reputation stems from the fact that the weasel is known to feed occasionally on the blood of its prey, rather than the flesh. Blood is a source of quick energy, and a weasel, possessing an extraordinarily high metabolic rate, makes good use of this source.

Perhaps the most distinctive characteristic of the short-tailed weasel is the manner in which it adapts to the northern climate's snowy winter. Every fall and winter in New England the weasel's entire coat is replaced. Over a period of three to five weeks, the brown summer coat is replaced by a white winter one. The reverse of this takes place every spring. The ability of the weasel to camouflage itself enhances both its ability to prey on unsuspecting mice, chipmunks, and shrews, as well as its ability to avoid detection by its own enemies.

We refer to the white short-tailed weasel in winter as an "ermine," and coats made of its fur during this season reflect this name. It is felt that the photoperiod, or length of daylight, and not the presence or absence of snow, triggers this remarkable change. Thus it is quite possible to see an ermine in its pristine coat of white before snow actually falls and after it has vanished in the spring. The short-tailed weasel's close relative, the long-tailed weasel, also possesses this remarkable adaptation, but it is not referred to as an ermine. [20]

THE FISHER

In Vermont, man has played a direct role in reestablishing the fisher as a valuable furbearer, just as he did in its near-demise years earlier. In 1957 the Vermont Legislature appropriated $5,000 for a porcupine control project, something at which the fisher is very adept. Between 1959 and 1967, 124 livetrapped Maine fishers were released in forty-one Vermont towns where extensive porcupine damage was evident. The purpose of the project was to introduce the one effective natural predator of the porky.

The porcupine has a penchant for chewing on practically everything he comes upon, so his lifestyle comes into direct conflict with that of man. Although best known for girdling, defoliating, and eventually killing softwood trees, he also has a distinct liking for axe handles, outhouse seats, tires, aluminum cookwear, plywood boats, camp floors, canoe paddles, and leather boots. The list is endless. A few porkies are not too bothersome, but when populations are high, damage to trees and other articles is significant.

How does the fisher kill a porcupine? This is a question often asked by those interested in outdoor things. The usual answer is that he flips the slow-moving porcupine onto its back and attacks the unprotected underbelly. However, Dr. Malcolm Coulter, a Maine wildlife researcher who has studied the fisher for many years, claims that it first attacks the porky's head, circling the animal and dashing in for repeated strikes. The weakened rodent is then flipped on its back, and the fisher rips open the quill-less belly to begin feeding.

The fisher so completely strips the hide of flesh and bone that it appears the skinning job might have been done with a knife. And oddly enough, although the fisher often ends up with quills embedded in various parts of its anatomy as a result of these encounters, the quills rarely cause a problem.

As expected, the fisher was extremely effective at controlling porcupines in Vermont. The fisher's range currently includes most of the mountainous regions of New York and New England. Some of the Great Lakes States also have a good fisher population, as do the southern provinces of Canada. [21]

THE WATER SHREW

At first sight, the furry three-inch water shrew appears quite harmless. Appearances, however, can be deceiving. This mouselike forest creature is in fact one of the fiercest and most aggressive predators in the animal kingdom.

Shrews are the smallest mammals in North America; they also may be the hungriest. A shrew eats from one-half to twice its body weight daily just to stay alive. With such a voracious appetite, shrews feed constantly, devouring all kinds of insects, as well as earthworms, snails, snakes, salamanders, mice, and even other shrews. As belligerent as they are hungry, shrews may attack, kill, and eat mice twice their own weight. Except for the duck-billed platypus, the short-tailed shrew—one of thirty species of shrews found in North America—is the only poisonous mammal in the world. It secretes a venom in its saliva, and a couple of well-placed bites will render a much heavier victim helpless in seconds.

A shrew eats as much as it does to fuel its high metabolic rate. Its heart beats seven hundred times a minute—up to twelve hundred when frightened—and it breathes ten times in the time it takes a person to inhale and exhale once. This high metabolic rate is due partly to a shrew's small size. All warm-blooded animals use most of the energy they take in to maintain a constant internal body temperature. Smaller animals expend more energy than large animals, and thus need to eat more to replace lost body heat. Some species of shrews will starve to death in seven hours without food. In fact, the creatures are popular among farmers because they consume so many insects and insect larva.

Shrews are rarely seen by humans. Active both day and night, and throughout the winter, they spend most of their time burrowing under forest litter or scurrying through tunnels created by other animals. Shrews have very small eyes, which are sometimes hidden in their fur, and their vision is poorly developed. They must rely on their acute senses of hearing and smell in order to locate prey. [22]

THE JAGUARUNDI

An endangered small wildcat, the jaguarundi, is eluding researchers attempting to document its presence in south Texas, at the northern edge of its range. They have not been able to livetrap and radio-tag even one jaguarundi, although they have tagged eight ocelots — the other small cat whose ecology and distribution they are studying.

Before the study began, this might have seemed surprising, because reports of jaguarundi sightings in south Texas were three or four times more common than those of ocelots, which occasionally have been shot or killed by cars. One of the researchers said they surveyed fifteen hundred trappers in twenty-five southern counties and found none who had caught or seen a jaguarundi.

The reported sightings probably have been of feral black cats, raccoons, or other similar-sized animals, Tewes says. The jaguarundi is the size of a large domestic cat, weighing about fifteen pounds, and is all one color, gray or reddish brown. However, it is long and low-slung, with a small, flattened head, and moves quickly, like a weasel.

Jaguarundis are secretive, live in heavy underbrush, and are found throughout Mexico, Central America, and parts of South America. Although the jaguarundi used to be considered common, four of the eight subspecies are now listed as endangered. The habitat of the form found in Texas (*Felis yagouaroundi cacomitli*) extends into Mexico. The same is true of another subspecies, *F.y. tolteca*, whose range reaches up into a tiny portion of Arizona. The other two endangered subspecies are Central American.

Only spotty observations of jaguarundis have ever been recorded, and scientists are not even sure whether the animal is solitary or lives in groups, let alone what its dietary needs are. Besides gathering this basic behavioral information, the researchers — based at Texas A & I University in Kingsville and federally funded — want to map any remaining populations and determine how much brush country habitat is left.

Of thirteen counties in south Texas, the researchers have found that much of the brushland has been cleared for pasture or cropland. [23]

THE COUGAR'S NEW CLOAK

For centuries, the cougar, *Felis concolor* (cat all of one color) has been blamed for crimes it never committed, ascribed traits it never possessed, and credited with feats it could never accomplish. The predator has been hated, maligned, persecuted, and feared. A negative image was as much a part of the beast as its long, black-tipped tail. But all that has changed. Today, the cougar wears a new cloak—a cloak of respectability.

The good image came hard. For nearly two centuries, no one had anything good to say about the cougar. Even Theodore Roosevelt called it a "big horse-killing cat, destroyer of the deer and lord of stealthy murder with a heart craven and cruel." Spurred on by grisly stories of livestock and wildlife predation, states paid up to $50 for each dead cat. Every missing lamb, every strayed dog, every late-for-supper child was thought to have been snatched and eaten by some marauding cougar. An aura of mystery—and hatred—grew to envelop the big feline. Homesteaders, pioneers, and stockmen shot the cat on sight. Poison was used to kill many more. Hunters and baying hounds roamed the woods in search of mountain lions.

As a result of the indiscriminate killing, along with habitat destruction, cougar populations plummeted. Where cover was scarce, such as on the prairie, the cats were quickly eliminated. By 1900, cougars had virtually disappeared from the eastern United States. Only in the vast expanse of the West did the cougar find refuge, but even there its days appeared to be numbered. And no one mourned.

But that was yesterday. The cougar of today has shucked its old image as neatly as a buck drops its antlers. Federal and state agencies now try to protect, not persecute, the mountain lion. Frequenters of wild places now thrill, not cringe, at the sight of the animal. North America's large cat finally has earned the public's admiration.

The object of this newfound affection is a solitary, secretive creature that is the epitome of successful carnivore evolution. Powerful legs drive the cougar's slender, supple body. Retractable, needlelike claws adorn its padded feet. The big cat's coat may be brown, red or gray. The ears are short and round, the teeth long and sharp. The powerful jaws can snap a deer's neck in an instant. Males average 160 pounds and females 135. From nose to tip of tail, the cougar may measure eight feet. It is the unquestioned lord of its environment.

The cougar—or mountain lion, puma, or panther, as it is called in different regions—can live in jungle, desert, prairie, or mountain. Its range once extended from British Columbia to Patagonia and from California to Florida, making it one of the most widely distributed mammals in the Americas. Today, cougars flourish in the United States only in the eleven western states and Texas, and there's a remnant population in Florida. Biologists are looking diligently for proof of its existence elsewhere in the East, and occasional unconfirmed sightings are made in states ranging from Arkansas to Minnesota.

Almost everywhere the cougar resides, humans now accept, even enjoy, its presence. What has changed, of course, is not the cougar, but humankind's perception of it. [24]

NATURE'S GOLDEN RACER: THE CHEETAH

The cheetah is the fastest wild animal on earth. From a crouching start, he can reach forty-five miles per hour in two seconds. A second or two more, and he's careening along at over a mile a minute, his long, thick tail acting as a counterbalance so he can corner like a jackrabbit. Zoologists estimate the cheetah's top speed to be a scorching seventy-five miles per hour.

One hundred-thirty pounds of rangy muscle (relatively lightweight compared to other big cats), the cheetah is clearly built for speed. Long-limbed and lean, he stretches more than seven feet from nose to tail tip and stands two and one-half feet high at the shoulder. His head is small and streamlined, with markings like tear-shaped ebony stains that run from the corners of his eyes to his mouth. His body is gracefully tapered from chest to waist, his coarse yellow fur punctuated with black dots.

There are many who confuse cheetahs with leopards. At first glance they look enough alike to be twins — both are tawny-furred, spotted cats. But beyond this superficial similarity, they are different.

The thickset leopard, heavier by thirty pounds, operates in the shadows of the night, relying on heavy cover, stealth and close-up ambush tactics. The trim cheetah is a creature of the day, who likes open spaces and long vistas. Its eyesight, like that of all cats, is keen, enabling him to pick out even well-camouflaged prey at great distances.

The cheetah is distinguished even further from the leopard and other felines by some distinctly uncatlike features. His legs are long and thin-boned, like a dog's. His head looks too small for his body, and his jaws and teeth are undersized. Even such feline basics as tree climbing do not always come easy to him.

For all his tremendous speed, the cheetah has one serious shortcoming: lack of endurance. He is a sprinter, not a long-distance runner. If an all-out chase lasts for more than about three hundred yards, he runs out of steam.

The cheetah seems almost painfully aware that he lacks the power, presence, and savagery of the other king-sized tooth-and-clawers. Except on rare occasions, he goes only after the smaller antelopes, killing quickly and cleanly. It is in keeping with his character that when he tries to roar, the sound that comes out is more like a meow. In moments of contentment he may even chirp like a bird. [25]

THE WILY, INDIGESTIBLE ARMADILLO

While wildlife all over the earth is diminishing, the strange-looking armadillo is not merely surviving but flourishing. What does the armadillo have that other animals don't?

A good suit of armor, for one thing. The armadillo, whose name comes from a Spanish word meaning "little armored one," is encased in a suit of hard, bony shells. One shell covers its head, and two shells connected by movable bands cover its body. The animal's tail is covered with hard rings that are locked together.

When taken by surprise, the armadillo curls up in a tight ball to protect its soft belly. Faced with such a jaw-breaking prospect, most would-be predators are willing to look elsewhere for their dinner.

Usually, the armadillo's keen sense of smell alerts it to any danger. Then, in a flash, the animal digs itself completely out of sight. Once hidden in a burrow, the armadillo arches its back and wedges itself in tightly. In this position, it becomes impossible for anything to pull the mammal loose, even though the end of its tail might be sticking above ground.

If an armadillo is near a river or other body of water when danger approaches, it jumps in, fills its stomach and intestine with gulp after gulp of air, and thus inflated, floats serenely to safety. Stranger yet, the animal sometimes just drops to the bottom and walks across.

These natural defenses help explain some of the armadillo's staying power, but other mammals have evolved equally effective survival techniques and still their range and numbers decline. The armadillo's secret, scientists agree, is that it is one animal that has been helped, not hindered, by civilization's encroachment on the wilderness.

As the lowlands are cleared and the forests cut to make way for people, an ideal environment is inadvertently created for the armadillo, which is happiest in cut-over and second-growth areas. The farmer's crops also lure the armadillo into new regions, for the armored animal is very fond of peanuts, cantaloupes, watermelons, and tomatoes.

With so much in its favor, it has taken less than a century for the armadillo to march far from its traditional home in Central and South America, into Texas, across the wide Mississippi, east all the way to Georgia, and as far north as Kansas, where only the threat of cold weather prevents it from invading farther. [26]

THE SEA OTTER

Although the smallest sea mammal, the sea otter is the largest member of the weasel family. Weighing only seventy-five to a hundred pounds, it is much lighter than a dolphin or a seal, but still huge compared to its cousins, the wolverine and skunk. Like its relatives, though, it has a fine fur coat, an asset that has endangered the otter's survival.

The sea otter's coat lacks an insulating fat layer. Instead, dense underfur, which must be kept clean for insulation and buoyancy, traps air next to the skin. Exposure to an oil slick means certain death for the sea otter.

Northwest exploration helped to deplete the sea otter's range of habitat. Today the sea otter lives along the coasts of Alaska and California to depths of about eighty feet. Rarely coming to shore, the sea otter hunts sea urchins, abalone, mussels, and crabs. When it catches one, it tucks the shelled creature into loose folds of skin and dives for a rock. Then, lying on its back in the water, the sea otter places the rock on its belly and smashes the shellfish against the rock. This tool-using behavior is one of several adaptions that have helped it survive.

The sea otter has strong webbed hind feet and a flattened tail, much like the beaver. But unlike the beaver, it swims on its back, propelled by its flipper feet and steered by its rudderlike tail. Its forepaws are clawed and help to hold food; its molars are flattened to crush shelled food.

Sea otters mate from age three on. Usually segregated by sex, the sea otters court and mate in the water, producing a single pup once a year. The pup is entirely dependent upon its mother for at least a year. She carries it on her chest and leaves it only to dive for food.

Now protected under the Endangered Species Act, the sea otter's survival is still threatened by food scarcity, violent storms, and oil slicks, not to mention illegal hunters. As weather and food have always been problems for the sea otter, man ought not to further threaten the animal but bear a large responsibility in seeing that it survives. [27]

ANIMALS' DIVING GEAR

Flippers, snorkles, goggles, wet suits: Humans will don some strange-looking equipment in order to feel more at home in the water. But no matter what we wear, we simply can't match the aquatic abilities of those animals who are born with special diving gear. These super swimmers include:

The dolphin—found in most of the world's oceans, these marine mammals are equipped with a sophisticated sonar system for locating food, obstacles, and each other in the dark water. The dolphin's eyes are covered with thick outer layers, with a gland that secretes an oily liquid to bathe the outside of the eyes. The layers and the liquid protect the dolphin's eyes so the salty sea water won't make them sting.

Although it must come to the surface to breathe, the

dolphin can store much more air in its lungs than a human can. When hunting for fish or squid, it can dive to depths of up to eight hundred feet.

The hippopotamus—how can the heavy, clumsy-looking hippopotamus be classified as a super swimmer? By holding in just the right amount of air, hippos become nearly weightless underwater. Then they rather gracefully half-swim and half-tiptoe along river bottoms.

The hippopotamus—the word means "river horse" in Greek—must spend most of the day in the water. Its skin dries out so fast that the big animal would die if it couldn't soak in water or mud. At night hippos emerge and graze on land.

The river otter—with its sleek, streamlined body and tail, a river otter can swim fast and turn quickly, and it can dive in a flash, leaving scarcely a ripple. Special flaps close automatically underwater, like hatches on a submarine.

Penguins—these highly specialized birds do fly, in a sense, underwater. Their wings are powerful flippers that propel them through the water swiftly. A penguin can swim as fast as twenty-two miles per hour. Penguin feathers make fantastic wet suits. Thick, short, rigid, and overlapping, the feathers and a layer of fat beneath the skin keep the birds from freezing to death in the icy water.

Some varieties of penguins spend months at sea, straying onshore only to breed, lay eggs, or molt. While diving for food—shrimp, squid, or krill—they usually surface every two or three minutes to breathe, although the Emperor penguin has been clocked at eighteen minutes underwater. This large penguin is also a deep diver; a vertical plunge of 885 feet has been recorded. That's almost twice as deep as the record for a human diver with scuba gear—and the penguin was just holding its breath. [28]

THE TRAGIC FACTS ABOUT WHALES

The killing of whales exemplifies, perhaps, mankind's darkest hour. These magnificent creatures, which have taken millions of years to evolve, are being mercilessly hunted and slaughtered by Japan, the Soviet Union, and a handful of other nations in defiance of a call for a worldwide whaling moratorium.

Modern whaling is big business. Huge convoys of ships roam the seas surrounding Antarctica searching for their prey. These fleets are equipped with sonar, helicopters, long-range explosive harpoons, and factory ships that can reduce an eighty-foot whale to a memory in less than one hour.

Whaling is not humane. Although searching for whales is highly sophisticated, the actual whale kill is barbaric. The whale is killed by a two hundred-pound, six-foot-long iron harpoon, shot from a 90-millimeter cannon. The harpoon head contains a time fuse grenade that literally blows the whale's entrails apart seconds after impact. The whale may spend hours in agony, and more than one harpoon may be necessary to bring death.

Whales are slaughtered to provide products for which there are substitutes. Whales are killed for animal feed, industrial oils, fertilizer, perfume, soap, shampoo, gelatine, and margarine, to name just a few uses. Inexpensive and plentiful substitutes exist for each of these whale by-products.

Whales are fascinating creatures whose existence has interested scientists, artists, and writers for centuries. Here are some interesting facts about whales:

The largest creatures on earth, whales are warm-blooded mammals. They are not fish and need to breathe air to live. Most are gentle, even playful, both among themselves and around man.

Whales travel in herds, often migrating year after year to the same areas.

Whales can communicate with each other through a series of high-pitched noises that sound like singing and can be heard in open waters more than two hundred miles away.

The blue whale is the largest of all whales:
Larger than thirty elephants
It weighs more than 2,000 people
Its heart weighs 1,200 pounds
Its tongue weighs one-third of a ton
Some arteries are so large a small child could crawl through them
A newborn calf weighs two tons and is twenty-five feet long

Whales normally cruise at about six knots — approximately twice as fast as a person usually walks. They are capable, in short bursts, of speeds up to fifteen to twenty knots and have been known to pace large ocean liners.

The brain of the sperm whale is perhaps the most complex brain ever evolved on earth. And there is no doubt that these complicated brains are used for intelligent, complex communications. [29]

THE BOWHEAD WHALE

The bowhead (or Greenland right) whale has a stocky, rotund body and weighs an average of twenty-five tons. The whale's head makes up one-third of its fifty to sixty foot length. Its huge mouth holds the longest baleen plates of any whale — up to fourteen feet. The bowhead's spout is V-shaped and may shoot as high as thirteen feet into the air. Their blowholes are located above the highest point of their skulls, which may be an adaptation necessary for breaking up through thick ice in order to breathe. Other bowhead characteristics suited for their cold surroundings are very thick blubber and the longest underwater endurance record of any whale. They have remained submerged for as long as one hour—all on a single breath.

The bowhead's range is limited to the Arctic. In winter and summer months, their location is not really known. However, in the spring and fall, the small bowhead population, numbering an estimated thirty-five hundred to four thousand individuals, migrates along the Arctic coast of Alaska. Breeding occurs in late summer, and the gestation period is believed to last about twelve months.

The bowhead's habits are little-known, making them the most mysterious of all whales. They travel singly or in pairs, often accompanied by beluga whales. They use their large mouths for the "skimming" method of feeding, which allows them to glide along the water's surface, mouth open, collecting plankton and small shellfish.

The bowheads are slow swimmers and easy to kill with cruder hunting methods, such as harpoons and small boats. Their buoyancy keeps them afloat after being killed, a great advantage for early whalers towing whales back to shore. These unfortunate traits caused the bowheads to be severely overhunted and nearly exterminated by 1900. Their name *Greenland right* is derived from the idea that they were the "right" whales to hunt. One bowhead yielded as much as twenty-five tons of oil and one and a half tons of whale bone to be used in the last century as corset stays, umbrella ribs, and skirt hoops. Bowheads are now protected under international law. [30]

WHITE WHALE

One seldom thinks of whales as milk white.

Oh, Moby Dick, perhaps, the great white sperm whale that Captain Ahab chased through the pages of an American literary classic. Moby Dick was a phantom, pure fiction, but at the mouth of the Saguenay River in Quebec one can be surrounded by milk white whales.

They are belugas, a name that in Russian means white whale. They are extremely wary animals.

Belugas are essentially Arctic whales distributed around the top of the world. They occasionally stray as far south as Cape Cod Bay or Long Island Sound in summer. But belugas are a regular feature of the St. Lawrence, where they are rather permanent around Tadoussac and may stray to Quebec City. They must feel at home in the Saguenay, for the August temperature of the stream is thirty-four degrees Fahrenheit.

Belugas are small whales averaging nine to sixteen feet long. They weigh up to fifteen hundred pounds. Their appearance is unusual, because they have small heads and appear to have necks. Belugas travel in pods, often traveling in groups exceeding one hundred animals.

Even in summer, belugas rise and blow in a sort of ballet pattern. In a group of five, never more than one is at the surface. In winter, this characteristic serves the animals well. They break a hole in thin ice, or keep open a hole in thicker ice, and then rise to blow in rotation, giving each member of the herd an equal chance for survival. Belugas can remain submerged only about fifteen minutes, so cooperation under Arctic ice is crucial.

On the St. Lawrence, visitors look for white whales and black whales, the latter classification including the general run of species. While watching the belugas, many are convinced that they are seeing black whales with them, because young belugas are dark brown through the first year. They are mottled the second year, yellowish the third, and turn white the fourth year. Their cousins in Hudson's Bay follow a slightly different color sequence, beginning life dark blue.

St. Lawrence belugas are wary, because until recently the animals were killed indiscriminately by Quebec fishermen. Belugas have little commercial value, although the animals are important to the more primitive Eskimo tribes. Fishermen have killed the animals because they eat fish, and fishermen can be upset about anything that eats fish without paying them. [31]

QUESTIONS FOR A BLUE WHALE

I am only three hundred feet away from you, my big friend, but it might as well be three hundred miles. I have been following you and yours for four years now — sometimes circling overhead for hours at a time, watching and wondering as you go about your daily business — but I still don't know very much about you. And neither do the few scientists who have been studying you for not so many more years.

Oh, we know all the superlatives — that your heart weighs half a ton and your tongue is as big as a taxicab, that 15,000 pints of blood course through your arteries and veins, that your newborn babies can gain as much as ten pounds an hour. But there is so much more we don't know, the things that might unravel the mystery that surrounds you in your deep blue seas.

Where do you come from, where do you go, where do you breed? Where do you sleep, do you sleep? How deep do you dive? I've watched you make ten-minute dives in a thousand feet of water; do you go to the bottom? If so, why? Your food is within a few hundred feet of the surface. Does the pressure scratch your itches? Or does it simply feel good? What natural enemies do you have? What diseases do you get, and are they fatal? How do you communicate, how do you navigate, how do you find food? How is it that you are found in both polar hemispheres but never crossing the tropical seas? And, forgive my seeming selfish and unfeeling, but how is it that not a single malignant tumor has been found in all the thousands of you that we have killed and cut up?

Perhaps mystery is your lot, and ignorance ours, in these matters. Perhaps it is the price we have to pay for having cared so little about you these past eight hundred years we have been hunting and killing you. Perhaps we had best not make direct communication, you and I, for once the barriers are breached, I don't know how we'll get past the burning question. Why? [32]

Birds

RITES OF SPRING

Many techniques are employed by creatures inhabiting this earth to insure the continuation of their species. Attraction of mates is accomplished in a variety of ways, from the colorful displays of many birds, fish, and reptiles to the deafening concerts of familiar amphibians and insects. All senses are utilized in this most instinctive of drives: Moths emit chemical pheromones, lightning bugs flash nocturnal messages, toads trill—ardor knows no bounds!

Mating rituals involving movement and often touch, whether on land, in the water, or in the sky, have triggered reproductive responses from females for generations. Males of many species dance, posture, call, and otherwise behave peculiarly and conspicuously immediately before mating. Often females join in, sometimes displaying in the same way, sometimes reacting to male displays with different ones of their own.

Northern New England boasts many inhabitants that participate in such courtship rituals, some more easily observed and thus better known than others. Among these is the American woodcock. Because it is such an inconspicuous bird during all other seasons of the year, the woodcock appears exceptionally noticeable during his spring sky dance. This performance commences with a nasal, insectlike *peent* note emitted every twenty seconds or so by the male bird on the ground. If close enough, one can detect a soft, gurglelike *tuko* note, uttered with bill closed, immediately before each *peent*. After as many as 160 *peents* (roughly five minutes), the male woodcock flushes and rises into the twilight sky. Fifty feet or so above the ground, his accelerating wing beats create a musical twitter. At first in wide circles (that encompass as many as three acres) he rises higher and higher in ever-smaller circles. The twittering stops as he momentarily appears to hover at the peak of his ascent (200 to 500 feet up), after which he proceeds to glide to the ground on a zigzag course, all the while uttering a series of liquid chirps. He lands close to the same spot from which he took off a minute before, and after repeating his ground peenting call for several minutes, he again launches himself skyward. This performance can extend throughout the dusk hours and well into a moonlit evening. Should his display succeed and a female woodcock be lured into his territory, the male walks stiff-leggedly with raised wings to the female, uttering only the *tuko* notes. Mating soon follows. [1]

GROWING UP IN A NEST

It's a wonder that any young birds survive to reach adulthood when one considers the length of brooding time, plus the number of days the nestlings have to be fed and watched over until they can fly. The fledgling time is particularly dangerous, since there exist so many creatures who depend on flesh for survival, and are on the alert for tasty eggs or a nestful of little birds.

Our Agnes was so fat and lazy that she wouldn't think of exerting herself by scrambling up a tree, even an easily climbed tree like our apples. But she did have a suitor in the neighbor's cat, and we heard reports of raccoons down the road. We knew there were snakes about because Katie started to pick up the hose one evening and it wriggled off.

Most of the birds that one hopes to attract are perching birds—termed passerines, from the enormous and successful (in the sense of evolution) order of Passeriformes—which comprise three-fifths of all the birds in the world. Swallows, wrens, thrushes, sparrows, larks, warblers, flycatchers are only a few of the passerines. Most of these birds are alike in that they don't begin to incubate until the correct number of eggs had been laid. How does a bird know what constitutes the "correct" number of eggs? It seems to have something to do with the feel of the right number of eggs in the nest. One poor flicker, whose nest was robbed daily of an egg, laid seventy-one eggs in seventy-three days. I hope the scientist-turned-egg-thief finally let her keep a few, but the experiment did indicate that such birds are correctly called "indeterminate layers." In contrast, "determinate layers," like the plover and the sandpiper, lay four eggs, and will not lay more to fill the nest if any eggs are taken.

The number of eggs laid depends, too, on the rate of attrition each species must overcome. The more dangerous the circumstances in which a bird is typically reared, the more eggs it lays. The hummingbird for instance, lays only two eggs in its tiny and well-concealed nest; but a duck or pheasant whose eggs are laid on the ground and whose young must avoid all manner of dangers might have as many as fifteen in a clutch. Whatever the mechanism may be that allows a female passerine to know that she has laid the correct number of eggs and that she must start the brooding process, it ensures that the eggs hatch at the same time and have an equal start in life. On the other hand, the raptors, those birds like owls and hawks that depend on flesh for food, start to incubate from the day the first egg is laid, to make sure that their young hatch at intervals. Then, if times are hard and there's a scarcity of food, the oldest owl nestling may eat her little brother. I imagine that this is another instance of the survival of species, that at least the elder sister will have something to eat. [2]

MASTERS OF ADAPTATION

When the first white men tramped through the woods where the traffic on New York's 42nd Street now roars, they reported that the song sparrows, vireos, and warblers created such a clamor that ". . . men can scarcely go through them for the whistling, the noise, the chattering."

Those songsters have long gone, departed from Manhattan, but birds have their ways of adapting if given a chance. The New York City of today harbors rock doves, night hawks, robins, chimney swifts, house sparrows, and starlings. Their nests are made of the twigs and leaves from the parks, the paper, string, and bits of material left about. Seeds, insects, berries, and discarded foods sustain them.

Some birds have positively gained by man's appearance on the scene. Even if the song sparrow, vireos, and warblers are seen no more along the forests of 42nd Street, the chances are, if you happen to steal along there on a summer evening, that you may hear a sharp pee-ik overhead. Because you are a bird-lover and because you have keen ears that catch the unusual noncity sound, you'll look up and see a night hawk sweeping overhead. Night hawks, who aren't hawks at all but members of the whippoorwill family, formerly had to make do with finding a barren, sandy, or rocky place in which to lay their eggs. Along came industrial man to put up flat-topped buildings by the thousands, and a flat-topped building was just what night hawks had been wanting ever since the first of them took to the air. To the birds the tar and pebble roofing looks just like their natural nesting sites; and best of all, their former predators, foxes, can't work elevators. The night hawks are thriving.

Barn swallows had to nest on ledges until white farmers came to build their barns all over the land. Indian lodges and tipis were obviously not suitable for swallow nests, but the dusky interiors of the high-roofed barns were exactly suited for the swallows' habit of nesting in colonies. Their numbers have increased rapidly.

Bank swallows think our road-cuts are splendid; and chimney swifts, who formerly nested in hollow trees, discovered the benefits of our chimneys. Now, alas, we have begun to tile our chimneys and the swifts are having a hard time of it, since the smooth tile affords no little ledges where they can attach their nests of glue and twigs. [3]

SPRING BIRD MIGRATION

Although many bird watchers look forward to March as the beginning of spring and bird migration, often it is not until April that both become reality. It is true that migration occurs in March, particularly among waterfowl, but also among such species as killdeer, woodcock, blackbirds, robins, and some sparrows. However, by mid April observers anywhere are likely to see a few true migrants. Among the earliest birds to arrive in the north in numbers are the tree swallows. Flocks are often seen coursing low over ponds and rivers on overcast or misty days. Another species that is encountered at this time is the phoebe. The phoebe tends to be solitary, often perching close to the ground on an exposed perch from which it sallies forth to secure its food. Although it is not much to look at, its feathers being a somber brown, the phoebe's confiding nature and tail-wagging habit make it a favorite of many nature observers.

The highlight of the spring migration comes with the waves of warblers that pass through during May, although April is not without representatives of this group. Both the yellow-rumped (myrtle) and palm warblers are familiar migrants in April. The palm warbler, with rusty cap and yellowish underparts, seems to have adopted the phoebe's curious tail-wagging habit. A rustling of leaves often announces the presence of palm warblers as they forage on the ground and low in the shrubbery. Yellow-rumped warblers are the most familiar of the late April migrants. Flocks can often be found in wetlands and along the margins of ponds.

Another common migrant during this period is the ruby-crowned kinglet, one of the smallest landbird migrants in New England. Its color is a rather uniform drab olive, and it has a distinct habit of flicking its wings. The most remarkable aspect of this tiny creature is its lengthy song, which it sings frequently during migration. The song, of about five seconds' duration, consists of several introductory high notes followed by several lower notes and concludes with a rolling chatter. The song is surprisingly loud for such a tiny bird.

An extremely predictable but highly variable phenomenon in April is the hawk migration. Early in the month, species such as the sharp-shinned hawk, turkey vulture, osprey, and American kestrel are migrating but are seldom seen in high concentrations. Hawkwatching aficionados wait until the latter part of April for the anticipated influx of broad-winged hawks. The passage of these birds is highly unpredictable and extremely dependent on weather conditions. Warm, balmy days with southwest winds often produce no hawks. Many feel that on days like these the hawks are soaring over so high that they are not visible. Then a cool northwest wind might produce a modest flight of low-flying birds. If migration were easy to predict, it would take some of the fun out of bird watching. These are just a few of the species that can be routinely expected on an April day. Certainly, others also can be found before the first warm spell in May opens the floodgates of migration and the long-awaited orioles, tanagers, warblers, and thrushes become a familiar sight. [4]

FEATHERS

In flight, a bird's life depends upon its feathers. Large birds may have as many as 25,000, each of which contributes to the streamlining of its body and each of which must from time to time be preened and kept in good condition. But no feathers are more important in the air than the strong, durable flight feathers of the wings and tail. One large, primary flight feather may consist of a thousand barbs and a million barbules. Each tiny component within a feather, each feather on a wing, each bone and muscle that lies beneath, all work together in an aerodynamic design, the efficiency and versatility of which are still the envy of the best aeronautical engineers.

The feather itself is at least 140 million years old. The earliest known fossil feather, perfectly preserved in fine-grained limestone, was found in a Bavarian quarry where lithographic stones were being carefully mined. The fossil was identical in every apparent respect to a modern-day feather. Significantly, the vane on one side of the quill was wider than the vane on the other side. The asymmetry is a feature of flight feathers only; it eventually would be recognized as certain proof that the owner of that ancient feather could fly. Soon afterward, a complete skeleton of this first feathered lizard burst upon the scientific world: *Archaeopteryz lithographica*, "ancient wing of lithographic stone".

Today, birds are one of the largest classes of vertebrates, exceeded only by fish. They comprise 8,700 species ranging in size from the condor, with its wingspan of ten feet, to the tiny hummingbird. By contrast, mammals, which evolved on a nearly identical time schedule, have given us only 4,000 present-day species.

Feathers serve two primary functions: thermoregulation and flight. The feathers on totally flightless birds, like the kiwi, ostrich, cassowary, and emu, are used mainly for thermoregulation and so have lost, for lack of need, the hooked barbules that make normal feathers aerodynamically sound. The resulting body covering of these birds looks and functions remarkably like the fur of mammals. Highly specialized divers such as the penguins, which also no longer fly, have developed a dense covering of hairlike body feathers, which like the fur coat of seals is well adapted to the frigid waters of Antarctica.

The origin of the feather, like that of God and the beginning of the universe, still eludes us. [5]

HOW BIRDS CHANGE THEIR CLOTHES

Have you ever found a feather on the ground—and wondered how the bird happened to lose it?

It could have been accidental, but more likely it was simply one of the feathers shed by the bird while molting. All adult birds go through at least one molting period each year, usually late in the summer following the nesting season. It's a gradual process lasting several weeks, sometimes months. Developing an entirely new set of feathers is also a time of considerable physical strain—so much so that many birds reduce their activities and all but cease singing.

The loss of feathers usually occurs in an orderly fashion, following a specific sequence for birds of the same species. Wing feathers, for example, are lost and replaced in such a way that the bird always is able to fly. Only waterfowl, such as ducks, geese, swans, and flamingos, are exceptions to this rule. They lose all their flight feathers at one time; there's no flying until the new feathers have grown out sufficiently to make flight possible.

Why do birds molt? For the same reason that you replace older clothing. In time, a bird's feathers become worn and the colors fade. Indeed, it is somewhat amazing that a bird's plumage lasts as long as it does. Feathers are slowly worn away, though, by contact with tree leaves and branches while flying and by rubbing against coarse grasses or dense shrubs when feeding the young. Sometimes they simply break off. And because, like human hair, a full-grown feather is a dead structure, it must be renewed. Molting is the process by which this is accomplished. Of course, if an entire feather is lost between molting periods, a new one grows immediately. This solves the problem of accidental losses.

Some birds have more than one molt per year (the ptarmigan—a Canadian game bird—has four). If there is a second molt, it occurs in the spring. With some birds, it involves another complete change of feathers, as in the case of the white-crowned sparrow. Others, such as the American goldfinch, may go through a partial change.

When you consider that most birds have at least 2,000 feathers (whistling swans have 25,000!), you can understand why these changes are a severe drain on their physical strength. It takes a lot more effort than dropping into a store for new clothes. [6]

MAN-MADE BIRD HOUSES

Small rectangular shelves put up in convenient places—such as under the eaves of a garage or shed—will attract robins and phoebes, who refuse to nest in an enclosed box. If you can't put the shelves under a ready-made overhang, put a little roof over them.

Here are some general recommendations for building or supplying nesting boxes:

When you're building an artificial nest, remember that

birds have specific preferences, so building an all-purpose nest will be just a waste of your time.

Be careful not to make the entrance hole too large.

Drill some small holes in the floor of the box for drainage.

Tin cans as nests can be death traps, because of the heat from the sun. Wood is the best material.

A few small slits or holes through the walls below the roof overhang will give better ventilation.

Guard the nesting birds from climbing cats or other predators by placing the boxes on metal posts, or putting metal guards on the posts.

The entrance should not be toward the prevailing wind.

Clean out old nesting materials before birds return in the spring. To make this task easier, the box should be made so that it can be opened easily.

Martin houses are painted white to reflect the sun, but dull colors are better for the outside finish for all other birds.

Birds are more apt to use a house if it is stationary and not swinging from a limb.

Gourds make economical bird houses.

New boxes should be set up in the fall to allow them to weather over the winter and be more acceptable in the spring. Also, nuthatches, downy woodpeckers, chickadees, and other hole-nesting birds will have a sheltering place in which to roost on cold stormy winter nights.

Use three-quarter to one-inch wood—cypress, white pine, cedar, yellow poplar. To make the box last longer, use brass screws, brass hinges, and galvanized or brass nails. Steel nails will rust. Sharp changes in temperature will cause nails to loosen and pull out, so screws are better. [7]

"MOBBING"

By its nature, the bird feeding station is a lively, sometimes quarrelsome, place where one may get the impression that it is every bird for itself. But, in spite of outward appearances, there is a surprising amount of community spirit that transcends the occasional fights and bickering. Birds warn each other of enemies through an easily understood, universal language. At times, birds also help each other in ways that seem selfless even by our standards. Evening grosbeaks at a Quebec feeding station, for example, treated an injured comrade as an outcast, but a "tenderhearted" male began taking care of the injured bird and saw to it that the bird had its turn at the feeder during slack periods. Another example was reported by New England ornithologist Edward Howe Forbush, who reported on an old, worn, partially blind blue jay that was fed and guarded by its companions.

Equally suggestive of a bond of brotherhood is the communal rite known as mobbing, where birds join together to thwart a common enemy. As defined by British ornithologist P.H.T. Hartley, mobbing is a demonstration made by a bird against a potential or supposed enemy belonging to another and more powerful species. It is initiated by a member of the weaker species and is not a reaction to an attack. Mobbing is conducted by more than one bird, and frequently every bird within sight and hearing joins in. From the smallest wrens to blustering jays, birds gather from all sides to deliver the strongest protest they are capable of. Amid a din of calls and chatterings, the frenzied birds, some flying, some moving from perch to perch, feign attacks and sometimes threaten an eye or other vulnerable part with their bills. Although birds have been known to kill snakes in mobbing attacks, usually these demonstrations are completely sham. The predator, perhaps ruffled, escapes with harm only to its dignity. After letting off steam and letting the predator know that it has been spotted and can no longer take its prey by surprise, the mobbing throng, having accomplished its mission, disperses.

That mobbing is an effective way to subdue a predator is obvious from the way the tables are turned. The cat or hawk that has been frustrated in its designs will, as likely as not, leave the scene. Birds can then resume feeding or whatever else they were doing. [8]

INDEPENDENT NESTLINGS

If the young of ground-nesting gamebirds, shorebirds, and waterfowl were as helpless as those of the familiar robins and sparrows, few would survive in their exposed situations. Instead they are bright-eyed, clothed in down, and able to scurry about shortly after hatching. Scientists have a name for this self-sufficiency — precocialism. Helpless nestlings are called altricial.

The differences between precocial and altricial birds encompass more than the nestlings themselves. The eggs of the former are larger than their altricial counterparts to accommodate the more completely developed embryo. Because the young leave the nest soon after hatching, the nests of precocial birds can be as simple as the tern's sparsely lined hollow in the sand or the ruffed grouse's leafy depression.

When the precocial bird's time has arrived, it cuts away the big end of the shell with the egg tooth on its bill and squirms free. Within a half hour it is standing on sturdy legs, eyes open, and downy duds dry and fluffy.

Because the nest and its discarded eggshells attract predators, the mother soon leads her brood from the scene. If an egg is tardy in hatching it is abandoned.

The young birds react to danger instinctively. Grouse, quail, and turkeys scatter and hide on command, their dead leaf camouflage rendering them practically invisible. Leggy birdlings like killdeer and stilts often opt to outrun their pursuers, especially when more than a few days old. Wild ducklings scamper across the water with amazing speed, hiding ashore or diving if closely pressed. Upland game birds soon learn to fly; grouse can flutter to low branches at the tender age of two weeks.

Even feeding is largely instinctive, although the hen encourages them to make the proper selections. Before long, brooding in cold or rainy weather is the only essential activity that remains for the devoted mother. [9]

THE UPSIDE-DOWN BIRD

Have you ever watched a nuthatch descend a tree headfirst? Next time, watch closely and you will discover how this agile bird is able to perform such an amazing act.

The secret is in how the nuthatch uses his feet. As he descends, he places one foot behind him, turning the claws backward to secure a firm grip on the bark. The other foot is kept forward under his breast, as a support.

Many people have speculated on why the nuthatch takes this headfirst route down a tree. The most plausible explanation is that it enables him to spot insects and larvae that go unnoticed by woodpeckers as they back down the same tree. If true, it's simply another of nature's ways of sharing what she has to offer.

Nuthatches are frequent visitors to winter feeding stations, and make friends easily. As one bird lover wrote: "They tame readily and come to my hand for food. When I am out in the woods, I have only to call and they will fly up and sit on my shoulder—waiting, of course, for the food that I always carry in my pocket."

In summer and fall, nuthatches consume a wide variety of insects—beetles, spiders, ants, flies, and larvae of tent caterpillars and gypsy moths. In winter, they can be attracted to feeders by sunflower seeds, acorns, peanuts, and other nut meats. But the surest—and most fascinating—way is to place a suet feeder on the trunk of a tree near your house. This will assure you a front-row seat for the antics of this upside-down bird.

By cultivating the company of a pair of nuthatches this coming winter, you will have the opportunity next spring to observe the showy courtship habits of the male. Seizing a piece of suet or a sunflower seed from your feeder, he will rush back to his waiting mate, spread his wings and tail feathers, and grandly present the food to her, as though it were a precious gift. He also serenades his chosen one—standing with his back to her and bowing stiffly between songs.

While the various white-breasted nuthatches are the most common, covering most of the United States, there are also the red-breasted nuthatch (actually more orange than red), the brown-headed and gray-headed nuthatches of the Southwest, the tiny pygmy nuthatch of the western states and several lesser-known subspecies. Whichever makes its home in your area—and they are all inclined to be fairly permanent residents—it's well worth the effort to attract a pair to your property. They're one of the stars of backyard birding! [10]

WHIP-POOR-WILLS

Few people have halfway feelings about the Whip-poor-will. Either they enjoy hearing it, or they can't stand its noise. In most cases they have become thus opinionated without ever seeing it.

A whip-poor-will is a goatsucker. How this designation came into being is hidden in the mists of time, for these camouflaged woods dwellers do not drink milk and have mouth parts unadapted to milking. They are insect eaters with very large mouths that are rimmed with bristles, which probably aid in preventing the escape of the insects it catches like a swallow, on the wing.

It makes no nest. The eggs are two in number, laid on the ground among dried leaves and so well camouflaged by color that a nest found one day may not be found the next. The young are not precocious, as are the young of many ground-nesters, and probably the protective coloration of the birds prevents the daytime discovery of the nest. The female leaves for feeding at dusk and the calls begin with gathering dark. A pair returns to the same breeding area year after year.

John Muir counted the number of continuous "whips" of a strident whip-poor-will and found that it "said its name" more than one thousand times without stopping. Shortly after this prolonged effort the bird began once more, and Mr. Muir was up in the hundreds again when, he reports, he must have fallen asleep.

Will it become an endangered species? Possibly, but how can a ground-nester be protected from egg-loving raccoons, foxes, and squirrels? All these predators are doing quite well, thank you. With a clutch of only two eggs, there is little to spare even to just keep the population stable.

Audubon said, "I have often heard hundreds in a woodland chorus, each trying to outdo the others. The fact that this bird may be heard at a distance of several hundred yards will enable you to conceive of the pleasure felt by every lover of nature who may hear it." [11]

BLUEBIRDS

One of the most rigid requirements for bluebird survival is a proper nest hole. The bluebird must have a cavity of some sort to even attempt nestbuilding, egg laying, and rearing young. Natural cavities in trees or old woodpecker holes have been their natural sites; now, nesting boxes provided by human friends are an important factor in the species' survival. Many of the natural sites have been taken over by the imported aliens who have wreaked such havoc in North America—the European starling and the house sparrow (or English sparrow), also a European import. The starling can be kept out of the man-made nesting boxes if the opening is exactly one and a half inches in diameter; they can't cram their bodies in, although if the opening is just an eighth of an inch larger, they can and will.

If the bluebirds succeed in nesting, they make model parents. The female lays three to five eggs in the grass nest, incubating them for about two weeks until the young hatch. Both parents feed the young, usually bringing food every five minutes until the babies are old enough to leave the nest fifteen to twenty days after hatching. Sometimes the young from the first nesting join in and help feed the second batch of babies. They stay together in loose family groups into the fall and probably migrate together to their wintering grounds in the southern United States.

In his book, *The Bluebird: How You Can Help its Fight for Survival*, Lawrence Zeleny lists five main reasons for this gentle bird's disastrous decline:

— Declining winter food supply: destruction of berry-bearing plants by man; hogging of available food by starlings.

— Adverse weather: bluebirds often freeze to death in unusually cold weather in the wintering grounds or on too early spring returns.

— Insecticide poisoning, particularly its heavy use in the orchards where they once nested.

— Destruction of habitat.

— Competition from alien birds, mainly house sparrows and starlings, who also nest in cavities.   [12]

CUCKOOS AREN'T CRAZY, JUST LAZY

The word *cuckoo* may be synonymous with "crazy" in American slang, but the European bird known as the cuckoo is far from crazy. In fact, the cuckoo is so clever that it tricks other birds into raising its young. Apparently convinced that parenthood is for the birds — other birds — the mother cuckoo lays her eggs in other birds' nests, then flies away forever.

That way, the cuckoos avoid the work of building nests and also the responsibilities of parenthood, such as feeding their young. It's a lazy, parasitic life, by human standards, but it seems to agree with the bird that is famous for its monotonous call and its appearance in Swiss wall clocks.

The European cuckoo is so clever that it doesn't pick just any bird to raise its offspring. While different varieties of European cuckoos lay eggs of different colors, each cuckoo lays her eggs only in the nests of birds whose eggs most nearly resemble her own. A cuckoo that lays bluish eggs lays them in nests of a warbler that lays bluish eggs. One whose eggs are speckled deposits them only in the nest of another bird that lays speckled eggs.

If there is another egg in the chosen nest, the female cuckoo carries it away or swallows it. So although she won't be around when her egg is hatched, the mother cuckoo plans for her offspring's future by making sure it won't need to compete for food.

In keeping with its mother's plan, a newly hatched cuckoo pushes any remaining eggs or other young birds out of the nest. It edges anything in the nest onto its back, and then rises up until the egg or young bird tumbles over the edge of the nest.

This ejection of all rivals is important to the young cuckoo, for it often grows to be much larger than its foster parents and needs all the food they would normally bring to their own brood of four or five.

What about the European cuckoo's American cousins? American cuckoos, at least, have better manners. They build their own rather flimsy nests of twigs in which they lay their own eggs. And they raise their own young. They also perform a valuable service because they feed on destructive tent caterpillars. Like their European counterparts, they aren't cuckoo. [13]

SWIFTS

In early May, the chimney swifts return to the north and take up residence in chimneys of all sizes. They still nest in hollow trees as their ancestors did before North America was settled. During migration, entire flocks of swifts can be seen going into large building chimneys at dusk. A large flock looks like a huge, living tornado.

Chimney swifts are not colorful birds, being entirely sooty black, but they do possess several interesting adaptations, and are among the most accomplished aerialists in the bird world. Their long tapered wings are in motion almost continuously, beating so fast that they seem to be moving alternately instead of together. If you think for a moment, you will recall that you have never seen a swift sitting on a wire like a swallow; the only place they land is inside their nesting or roosting site. Some species of swifts have even been observed sleeping on the wing; they do this by climbing to an altitude of several thousand feet, where there are no obstacles to worry about.

Because they spend so much time in the air, swifts' legs and feet are reduced to a very small size. Their claws are curved and make excellent hooks for holding onto the walls of their nest and climbing out. Chimney swifts apparently also use their feet to break off small twigs that they glue together with saliva to make their nest. Perhaps the most interesting adaptation of chimney swifts is their bristle-tipped tail. The central shaft of each tail feather extends beyond the side vanes about one-eighth of an inch, forming two strong sharp points that help to support the swift while it is at rest.

The diet of the chimney swift is composed entirely of insects, particularly beetles, flies, and ants. Their bill is adapted for capturing insects in midair; it is short but quite wide, rimmed with bristlelike feathers to form an insect net.

Enjoy the swifts while you see them, busily reducing the insect population as they twitter through the skies, for by early September they will be headed back to their wintering grounds in South America. [14]

THE PINE SISKIN

Have you ever had pine siskins at your feeders? You may have without knowing it, for this small member of the finch family is frequently mistaken for the female house finch, and sometimes for the goldfinch. The fact that it often feeds with these other finches adds to the confusion.

Three good ways to make positive identification are to remember that the pine siskin is about an inch smaller than the house finch and has heavier streaking and a sharper bill. There's also a tiny touch of yellow on the wings and near the end of the notched tail—but this takes a sharp eye to detect.

One thing you can be sure of: if these energetic little birds are in your area—and they roam over most of the United States in the winter—you can attract them with thistle or sunflower seed. They will also respond to white millet, canary seed, peanut hearts, cracked corn, and suet mixtures. But like their goldfinch cousins, their favorite food is thistle. Another similarity is that they have no preference in feeders, patronizing both the hanging variety and window feeders as well as low platforms and even the ground.

Pine siskins are remarkably easy to tame. As one bird lover in Massachusetts reported, "In a short time the birds came to regard me as a friend. Whenever I stepped outside, they would settle on my head, shoulders, and arms, and peer about for the food they had learned I held. Soon, one after another would come right into my kitchen looking for the handful of seeds that they knew would be forthcoming. Now and then members of the flock would spend the night in the warm room, sleeping on the clothesline that I stretched just below the ceiling."

Like many other species, siskins tend to travel in flocks during the winter, sometimes with as many as two hundred or more in a group. As spring approaches, these flocks break down into smaller ones, then into groups of a few birds, and finally into pairs. By this time, most siskins have returned to their breeding grounds in Canada and the northern reaches of the United States, although some remain in mountainous regions of the South and Southwest.

Nesting occurs chiefly in evergreen forests. Sometimes pairs are isolated, but often there are colonies with nests only a few yards apart. Despite their name, the birds do not insist on pine trees, also nesting in such other conifers as fir, spruce, cedar, and cypress. [15]

GROUSE

Throughout April and much of May, I frequently heard the low, motorlike sound of the ruffed grouse, or partridge, coming from one corner of our property. Not until that time did I realize what a perfect habitat the land was for grouse. About forty years ago the area, then covered with a white pine stand, was the object of an apparently halfhearted logging operation. Today half the property is covered with hundred-foot white pine (many the thick-trunked, much-branched "bull" pine) and with hemlock; the other half is covered with poplar and white, black, and yellow birch, with some beech, oak, and maple mixed in. Large rotting pine logs lie askew amid the broad-leaved trees. Ferns and brambles have filled in the sites where the sun reaches the forest floor. The land provides abundant and varied food and protective cover for grouse throughout the seasons.

Grouse are primarily ground dwellers and require not only protection from ground and aerial predators, but also suitable sites for the male to "drum" and display himself. The large decaying logs left during the logging operation provide perfect stages for this purpose. The display, or strutting, behavior acts as a visual means for communicating dominance and is done in the presence of another grouse. The male will swell its breast, ruff up the area around its neck, make its tail feathers erect, droop its wings, and lower its head. He then begins strutting, almost running, toward the other grouse, hissing and shaking his head. Drumming serves as an audible advertisement of the male grouse's presence, for the purpose of attracting a female and marking a territory.

The male grouse has a favored, or primary log, as a drumming stage. He usually goes to the large end of the log and the same point each time. With tail down over one side, and claws clutching in, he begins the drumming slowly at first, then with increasing tempo until the final weak thump. The first drum is made with a downward and inward motion of the wings, feathers close to the body and wings drawn back. Instantly the forward stroke follows.

The drumming, or thumping sound, is caused by the forward and upward strokes of the curved wings compressing the air with much vigor. During this ever-quickening process, the tail becomes more and more flattened against the log, allowing the bird to maintain its balance. The last thump is a weak-sounding one, made by a forward and downward wing movement. Once the drumming has ceased, the bird pitches slightly forward, and the tail lifts from the log as if it were a compressed spring.

While I have yet to actually see the male performing, I have a clear vision of the speed and vigor of the drumming and hope to get a firsthand look the next time I hear the muffled "lawnmower" starting! [16]

RAISING A GREAT HORNED OWL

In the stillness of a winter evening in late February, two tiny, down-covered young of a great horned owl emerge from their shells. Their eggs, which were laid one month earlier, have been kept warm by a maternal cloak of feathers through several days of subzero temperatures, through rain, sleet, and snow. The time to grow for these young birds will be during some of the harshest weather the season will bring.

The owl's nest was originally built by a red-shouldered hawk. Saddling the base of a large branch forty feet high in a great white pine, the nest is made of pine sticks and dead needles and lined with buff-colored down from the mother owl's breast.

But the young great horned owls will not see their home until their eyes open around ten days after birth. Here they will be fed for six to seven weeks until leaving the nest. The magic of flight and self-reliance will come when they're ten to twelve weeks old, as the warm June sun is upon them.

Why would any bird choose such a dubious season as midwinter to raise its young? What advantages could possibly exist that outweigh the hardships of the winter season—the difficulty of finding prey under a deep blanket of snow or of keeping young birds warm in the blowing winter winds? Perhaps there is less competition for food for their young, for great horned owls are first in the season to raise their brood. Or maybe what food can be found is easier to spot when the leaves are down. The early nesting habit of these owls is one more of nature's seeming eccentricities that gives us cause to ponder. [17]

OSTRICHES

The African ostrich, a flightless bird, may look like a clumsy, defenseless creature — an inviting target for attack by other wild animals. But with its great height and weight (it grows up to eight feet and may weigh 375 pounds) and its breathtaking speed (it can outrun a lion), it is far from helpless. In fact, it doesn't need working wings to escape its enemies, because it can outrun, outkick, and outsmart most of them.

The world's largest bird has several methods of escaping predators, but its best defenses are its legs and feet, which are engineered for speed. The ostrich has only two toes (all other birds have three or four), so its walk looks like a clownish hobble. When the bird breaks into a run, however, it surges gracefully along in seven-to-nine-yard stretches, concentrating the thrust of its feet onto its two toes.

With a top speed of fifty miles per hour, the only predator the ostrich can't outrun is the cheetah, but at the same time it can easily outdistance that spotted cat. And should any enemy get too close, a well-aimed kick from the powerful ostrich can be lethal.

One thing ostriches do not do is bury their heads in the sand to avoid danger. Presumably, this myth arose because of the female ostrich's habit of shielding her eggs by stretching her neck flat along the ground in front of her nest.

Considering the ostrich's brains and brawn, you might expect to find a large population of the big birds roaming their native African savannahs. In fact, ostriches once ranged across all of Africa and into the Middle East. But that was centuries ago. The birds were killed for their plumage, and their arid habitat was converted into irrigated cropland. By the 1850s the ostrich was already just about gone from the Mideastern deserts. The very last one is believed to have disappeared into a Syrian cooking pot some thirty-five years ago. Today the bird is confined to Africa's drier regions south of the Sahara, and development is quickly breaking up its habitat there.

Even with sound conservation efforts, the wild ostrich is likely to continue fading away, along with the rest of Africa's wildlife, because as fast as it can run, there's little the flightless ostrich can do to withstand the impact of civilization. [18]

THE BALD EAGLE

There is some good environmental news about our national symbol, the bald eagle. In the Chilkat Valley in Alaska in October, November, and December some 3,000 eagles gather along several miles of the river to feed on the dead and dying salmon that have come up the river to spawn. The river is one of the finest salmon areas because of its heavily timbered watershed, a neatly balanced ecosystem. But in 1979 the ecosystem was threatened when the state of Alaska and a lumber company signed a fifteen year timber sale contract. The timber to be clear-cut was mostly in the Chilkat Valley. The threat to the river, the salmon, and the eagles brought to a head a battle that had raged between conservationists and developers in the village of Haines, which lies near the mouth of the Chilkat.

The village, which had depended on the lumber industry, was severely depressed economically, hoping that the sale would revive its economy. But enough concern was voiced by natural resources councils, and by people from all over the country who feared for the eagles, that a proposed national wildlife refuge for the birds was to be established in the valley. That frightened many residents of Haines, who felt that a federal landlord would be much less responsive to their needs than a state one. Thus was the stage set for achieving an Alaskan solution to the Chilkat controversy. The governor declared a moratorium on logging in the valley and funded a cooperative resource study advisory committee to gather facts necessary to make land and resource decisions involving soil, timber, water, fish, and eagles. He commented, "If ever there were a seeming exercise in futility, it was when loggers, fishermen, miners, the chamber of commerce, and the conservationists were battling it out over the Haines bald eagle–logging issue."

But in February 1982 a bill was signed creating an Alaska Chilkat Bald Eagle Preserve and an adjoining Haines State Forest Resource Management Area. Among those signing were the Borough of Haines, the Audubon Society, and the lumber company. It can only be called the Chilkat Miracle. [19]

CONDORS

Radio beams from tiny transmitters attached to two freeflying California condors are guiding scientists to a wealth of new information that may help save this critically endangered species from extinction.

The remaining California condors—probably fewer

than twenty-five—range over thousands of square miles of rugged mountains passable only by a limited network of foot trails and dirt roads. Radio-tracking from an airplane, in conjunction with a ground crew in a four-wheel-drive vehicle, has made it possible to track the two birds' daily movements and study their habits. Already this has turned up at least twenty roosting and feeding sites hitherto undiscovered in years of study by observers on foot, and there also have been some surprises about the extent of birds' movements.

"All we could possibly have expected from the radio-tracking program is coming true," declared John Ogden, National Audubon's senior staff scientist at the Condor Research Center in Ventura, California.

The condor is North America's largest bird. It has a wingspread wider than a bus. A panel of scientists for the society and the American Ornithologists Union reported that without a drastic, now-or-never program of human intervention, there was no hope the species could be saved. The recovery program developed from that study. It calls for greatly expanded research, habitat protection, and captive propagation—taking some condors from nature, breeding them in captivity, and returning progeny to the wild.

The radio-tracking gear is providing vital information about what habitat must be protected if the condors are to be saved, and it is speeding up the collection of data about their habitats and movements. Among the surprises: The first radio-equipped bird to be released, an immature, cruised slowly up to the northern end of the condor range, stayed put for two months, then suddenly—for no discernible reason—flew 125 miles south in two days.

Ogden was worried that the radio-equipped birds would be bothered by the plane tailing them. They weren't. He found also that the radio signals can be picked up clearly seventy-five miles away. He can follow the birds in the air by plane, and when one lands can radio the position to the ground crew. The scientists on the ground can then drive and hike—guided by the radio signal—to a point close enough to the roosting or feeding bird to watch it by telescope. [20]

TURKEY VULTURES

As we canoed down the Connecticut River, we spotted two turkey vultures in a dead tree spreading their wings to dry.

The passage both in spring and autumn of turkey vultures now seems a regular occurrence in New England. Thirty years ago, this was not so. In fact, it is only within the last ten years that the huge blackish birds have been a regular migration feature.

Edward H. Forbush, in his 1927 volume on the birds of Massachusetts and New England, described the turkey vulture as: "size of smallish eagle; large blackish bird with long and rather broad wings." It was a good description of the bird, especially as we seldom see it on the ground in New England. As the bird was practically unknown in the region in Forbush's day, his accurate description of its aerial prowess probably came from observation of southern vultures. He said, "when once in the air and gaining height, it moves with the ease of a master. No other American bird is so generally celebrated for its perfect conquest of the aerial currents."

Forbush mentions the once well-known fact that the Wright brothers spent hours studying turkey vultures in their search for the secret of flight.

Audubon, in his *Ornithological Biographies* published in the 1840s, noted that the turkey vulture "is far from known throughout the United States, for it has never been seen farther eastward than the confines of New Jersey." Audubon thought of eastward in the same sense that made Maine downeast. "None," said Audubon, "I believe has been observed in New York, and on asking about it in Massachusetts and Maine...I found none knew it. On my late northern journey, I nowhere saw it."

In recent years, turkey vultures have been spreading into Canada. It is a visitor to Quebec, and a straggler to Newfoundland and Labrador, but a now regular breeder as far east as southern Ontario. William Mansell in his

North American Birds of Prey comments upon arrival of turkey vultures in southern Canada, where, as he notes, it seems to be a comparative newcomer. These birds may account for increased observations of vultures migrating over New England. [21]

KINGFISHERS

It's wonderful to have the kingfishers back. Pause by a pond or a stream, and you may hear an angry-sounding rattling chatter from above your head or across the way. A belted kingfisher is letting you and his fellow belted kingfishers know about his hunting territory.

I am always cheered to hear that irritated scolding, and delighted when I spot its maker, perched on a branch over the water. There is something at once pristine and prehistoric about the kingfisher — pristine because the white belt around its neck is such a pure white against the deep blue gray of its wings and back, and prehistoric because of the outsized black beak, the large somewhat flattened head with inconspicuous crest, and the shortened, almost tail-less body on squat little legs. Males and females are similar in size (eleven to fourteen inches) and look alike except for the female's cinnamon brown belly band.

Fish, as its name implies, are the chief food for a kingfisher. Large wings enable it to swiftly scout the streams or pond for food, and to hover in place while searching below. Its head-heavy body is well adapted for diving and its long heavy beak for clamping closed on its prey. Once the kingfisher has caught a fish (usually under six inches long), it returns to its perch, bangs the fish on a hard surface until it stops moving, then tosses it or adjusts it in its beak so it can swallow it headfirst to avoid the prickly fins. The kingfisher swallows the fish whole and later regurgitates a pellet composed of the indigestible scales and bones.

If you've never noticed a kingfisher's nest, it's not surprising. The parent birds excavate a two to five foot tunnel in a sandy bank and build a nest chamber at the far end of it. Their toes are adapted to dig this tunnel with two of the front-facing toes joined together to make a powerful fork-scraper tool.

An average of six to seven eggs are laid, and the young are born twenty-three days later; although they are protected from the weather deep inside their sandy bank, they huddle together in a heap for warmth, because the mother, with her short legs, is unable to cover them adequately. The fledgling birds are ready to leave the nest in about four weeks but depend on their parents' help to catch food for some weeks after that.

The belted kingfisher is not an uncommon bird, but it is certainly a special bird, well worth looking for and watching. [22]

THE COMMON LOON

No summer is complete without the sound and the sight of the common loon, that remarkable bird whose laughing cry has given rise to the expression "crazy as a loon." A modern symbol of the north country, this primitive bird was revered by American Indians, who honored it variously as "The Spirit of the Northern Waters," "The Most Handsome of Birds," and "The Omen of Death."

Soon after the ice leaves the lakes, the loon returns from oceanic wintering grounds as far south as the Gulf of Mexico and immediately establishes a one hundred to three hundred-acre territory, which it actively defends from other loons. This is one reason a substantial-sized lake may become home for only one pair of loons. There are other influential factors. A loon becomes airborn from the water and needs a long runway to generate sufficient propulsion. Besides eating vegetation, frogs, and mollusks, a loon feeds on white and yellow perch, bluegill, and various minnow species. These flight and food requirements demand a large, relatively secluded lake.

Within a few weeks of its return, the loon begins building a nest. Because a loon can propel itself on land only by pushing forward on its belly in an awkward manner, the bird builds its nest very close to the water. The nest itself may be the top of an unused muskrat house, simply a depression on a secluded sandy or leafy beach, or an arrangement of leaves, sticks, mud, and grass.

The female loon lays one or two chocolate spotted olive-colored eggs about the size of goose eggs. In twenty-eight days the eggs hatch to reveal chicks that can swim within twenty-four hours. During the first few weeks the chicks are especially vulnerable and often ride on the parents' back for rest and for warmth. Around the second week in July, the young are most visible swimming around with their parents.

When observing loons, remember to keep a good distance away so as not to disturb them or disrupt their activities. Do not touch any eggs. If you observe other people bothering loons, speak to them and encourage them to respect these magnificent birds. [23]

THE BITTERN

Although ducks, geese, and swans are among the most celebrated residents of our dwindling wetlands, perhaps the most unusual denizen of the tules is the common but little-known bittern. This reclusive bundle of feathers, a member of the heron family, is far more often heard than seen.

Some liken the bittern's call to the sound of a stake being driven into wet ground; others claim it resembles the noise a hand pump makes when it is being primed. Whatever the description, the bittern's unusual vocal abilities have earned it the nickname "thunderpumper." The bittern's pumping, which serves as a warning to rivals and as a mating call, is caused by the bird's ingestion of large quantities of air, subsequently ejected in great gulps. Ornithology texts describe the call as *oong ka choonk, oong ka choonk*, but this does little justice to the thunderpumper's resounding notes.

The American bittern ranges throughout North America and is common to most areas in the United States, but only serious wildlife observers and nature enthusiasts generally know the bird. The tawny brown, medium-sized heron (twenty-six to twenty-eight inches tall) is most readily identified by its black wingtips.

Like other members of the heron family, the bittern haunts marshy swamplands and is also found along rivers and lakes. But the bittern does not nest in colonies and thus is rarely seen in large numbers. Rather, this ghost of the wetlands lives in solitude except during the mating period. It feeds on the familiar creatures of the swamp, crayfish, frogs, and small fish. The thunderpumper sometimes impales its victims on its daggerlike bill before swallowing them whole. So strong is the bittern's three-inch bill that Indian tribes used them for arrowheads. When the bittern mates, it constructs a well-concealed reed nest deep in the cattails, away from the probing eyes of hawks, minks, and water snakes.

The bittern's remarkable camouflage technique is another reason why this shy bird remains unknown. When threatened, the thunderpumper points its bill skyward, displaying a vertically striped brown-and-white breast. Remaining motionless, the timorous bittern blends in with surrounding reeds and vegetation. As the bittern's alert yellow eyes are set extremely low on its head, this allows the bird to observe the cause of alarm, even though its head is pointing straight into the air. [24]

WHOOPING CRANES

A few years ago in New Mexico we heard that some whooping cranes had just flown in from Alberta to the wildlife refuge at Bosque del Apache near Albuquerque. Sure enough, sitting in our car, for visitors are not allowed to disturb the birds, we could easily see the whoopers, who were with their foster parents, the sandhill cranes. The young cranes were part of the project undertaken by the government to increase, by man's help, the diminishing population of the scarce birds.

Whooping cranes lay two eggs, one of which may be neglected, so the idea was to take one of the two and give it to their cousins, the smaller and greyer sandhill cranes, to raise. The two species do not mix in the wild, and it was hoped that the whoopers would grow up and join their own kind at breeding and migrating time. The sandhills seemed to have done a good job of foster parenting, for there were their adopted children, already as large as they and much whiter.

The whooping crane is the tallest of all North American birds, standing more than five feet tall, and when full grown one can weigh as much as twenty-five pounds. Their body plumage, in contrast to that of their cousins, is dazzling white with jet black wingtips, and their bare head is bright red. The voice of the whooping crane, a loud, clear buglelike note, must have been one of the most thrilling sounds of the prairies years ago.

The birds have been hard put to adapt to the change in their habitat brought on by western man's invasion of their territory, draining the swamps that were their breeding grounds and shooting the huge birds as they made their long migration to and from Canada to the marshes of the southern United States where they wintered. Only after years of work by the Audubon Society and the United States Fish and Wildlife Service was the attitude of the public aroused to the point of concern about the loss of the whoopers along the migration route. But because the pressure of human population continues on their breeding grounds, the fate of these magnificent birds still is not certain. We would all be diminished by the disappearance of the whooping cranes. [25]

GULLS

If, along with its superb flying power and eyesight, and its cast-iron stomach, the gull also had a brain, I might be an admirer. Unfortunately, it does not. Some observers point out how "canny" a gull is, dropping clams and crabs and other shellfish on rocks and highways to break open the shells. But naturalist Niko Tinberger points out that herring gulls also drop shellfish on beaches and mud flats, not knowing the difference between a soft or hard surface. Other scientists have offered herring gulls synthetic shellfish substitutes, from wooden eggs to metal objects. The birds dropped them all, as they would shellfish.

Perhaps the gull doesn't have much above the beak, but it does have other more obvious assets. Its long tapered wings, streamlined body, and habit of straightening its feet back in flight to reduce wind drag enable it to soar for hours or to fly over great distances. It can walk, fly, and swim extremely well—abilities most birds do not have in combination. Most gulls are also "edgers," able to live where land and water meet, thus finding food in two environments.

Gulls communicate with each other by voice. They have specific calls, from screams to mutters, that refer to almost every activity: breeding, alarm, hunger, attack, begging, challenging, fighting, and assembling the flock. They have double-down and thick, well-oiled plumage that makes it possible for them to live in the coldest climate (or warmest) and spend long periods on the water without becoming wet and bedraggled.

One of the gull's most impressive assets is its ability to drink either fresh or saltwater. Although the kidneys of most marine animals and birds—even fish, for that matter—are not equipped to extract and excrete salt from the blood, special glands do this for the gull. Another pair of glands above the eyes strains excess salt through openings in the bill. Scientists from Brown University, studying the growth of simple organs with a single function, removed the salt glands from several herring gulls, placing them on salt-free, low-salt, and high-salt diets. They found that the glands of the gulls drinking only fresh water grew very little, but those with the high-salt intake quickly developed large protective glands to flush the salt from their systems.

In addition to all this, the herring gull is just plain long-lived. In 1967, Olin Pettingill found a herring gull banded in Maine that proved to be thirty-five years, eleven months, and twenty-two days old, at that time the oldest banded wild bird on record. Ten years is normal, fifteen not abnormal, and captive gulls have lived to forty-nine, one female laying eggs once a year for forty-two years.

True to their nature, young herring gulls are ravenous

from hatching. As soon as the two or three eggs in a clutch are pipped and the spotted downy chicks emerge, they become the center of attention for both parents. The adults feed their offspring many times a day, regurgitating partially digested food, even pecking this into small manageable pieces.

Apparently, the parents can recognize their chicks in a gull colony, but chicks cannot tell one adult from another. Parents have to be on constant guard against predators of their offspring, which cannot fly until six weeks of age. But in two months the young can use their wings well enough to escape most enemies. From that thumb-sized chick with the big mouth, the gull grows to an adult in about two years, weighing just under three pounds with its stomach empty (which it rarely is), and attaining a wing spread of five feet. [26]

PUFFINS

Years ago, Eastern Egg Rock Island in Maine was the spring and summer home of hundreds of puffins busy raising families. The island was soon discovered by humans, who killed the puffins, stole their eggs, plucked puffin feathers for pillows and mattresses, and took their wings to decorate hats. Some people ate puffin meat and used parts of their bodies as fishing bait.

By 1902, only one pair of puffins was left.

Thanks to the efforts of Stephen Kress and the National Audubon Society's puffin project, puffins are returning to nest and raise their young on this Maine island after nearly a century of absence. Kress hopes to spread the puffin population around so that if they become endangered in one place, there will be plentiful puffins elsewhere.

In order to repopulate Eastern Egg Rock, Kress collected ten-day-old puffin chicks from Newfoundland and brought them to Maine. Now a surrogate father, he dug little underground burrows for the chicks and fed them raw fish twice a day, just like parent puffins do. He didn't handle the chicks any more than necessary or talk too much around them, wanting them to grow up like wild puffins.

Wild puffins spend most of their lives on the open ocean. When they are ready to raise families, they return to their birthplace. Kress had to trick his transported chicks into thinking that Eastern Egg Rock was their real home so they would return there to nest.

One particular bird, Puffin Number 54, adapted very well to her new home. She stretched her wings, dug into her little house, and came out at night to exercise while her enemies were asleep. When she was fifty days old, she left her burrow for good, swimming into the night to begin four years of growing up. While she was away, Kress moved more puffin chicks to the island in the hope they would return to nest, as he hoped Puffin 54 would.

On July 4, 1981, his wish came true. Puffin 54 flew into sight with a beak full of fish and slipped into a crack in the rocks. She came out without the fish—a sure sign that she had a chick to feed. After almost one hundred years, the first baby puffin has been born on the island. During the summer of 1982, eleven pairs of puffins were nesting on the island, giving Eastern Egg Rock a permanent population of puffins. [27]

KILLDEER

The killdeer is a shorebird, related to the plovers and sandpipers, but it is often found feeding and nesting a considerable distance from water. Killdeer are about robin-sized, but with longer legs. Their distinctive plumage consists of a brown back and white belly, with two prominent black bands around the neck. In flight, a killdeer also shows a long white stripe in the wing and an orange rump. Its name is derived from its most familiar call, a loud, repeated *killdee, killdee*.

The killdeer is probably the most familiar and widely distributed shorebird, nesting over almost the entire North American continent. In the late 1800s and early 1900s, however, hunters almost eliminated killdeer populations in many areas. With protection, the killdeer made a successful recovery, and it has been recognized since as a very beneficial bird, as it eats great quantities of insects, particularly weevils, grasshoppers, and other agricultural pests.

Flat open areas that provide a clear view for the incubating bird are the preferred nest sites of killdeer. Bare gravelly patches in meadows or newly plowed fields satisfy that requirement. The four eggs are laid in a shallow depression that is sometimes lined with small pebbles or bits of plant material. The eggs are superbly camouflaged by a pattern of brown and gray spots and blotches. They are practically invisible to a person looking directly at them from a distance of only a few feet.

The young killdeer leave the nest as soon as their down has dried, following their parents and finding their own food. If danger threatens, they squat down on the ground and remain motionless, while their parents put on their famous broken-wing act. The attitude most frequently assumed in this ruse is as follows: One wing is held extended over the back, the other beats wildly in the dust, the tail feathers are spread, and the bird lies flat on the ground, constantly giving a wild alarm note. This performance continues until the observer comes very near, when the bird rises and runs along the ground in a normal manner, or at most with one wing dragging slightly, as long as pursuit is continued. [28]

A GUIDE FOR GANNET GAZERS

Bonaventure Island off the tip of Quebec's Gaspé Peninsula is the largest gannetry in North America, but unlike many remote areas where wildlife is viewed elsewhere in the world, it is easy to get to and well-equipped to handle an influx of tourists. To visit Bonaventure, you simply drive to the tiny village of Percé, then board a boat for a quick crossing to the island colony.

The northern gannet, one of the world's three gannet species, is compelling to watch. The birds are affectionate to their mates and belligerent to their neighbors. When a bird lands in the colony, its mate greets it with what seems like the passion of a long lost lover. The greeting begins with both birds pointing their bills toward the sky and then entwining their necks and heads. This is followed by much caressing of bills, heads, necks, and even wings. All this display does not sit well with the neighbors, however. Ruffled by the intrusion, they often peck the new arrival with their massive bills.

When a bird leaves the colony, it has a major problem to contend with before it can get airborne. Unlike most birds, which are built to take off from a standing position, the gannets, weighing six to seven pounds, must hurl themselves off the edge of the cliff. To get to the takeoff point, they must select the least hazardous route through the masses of pecking, flapping, wing-beating neighbors. Once at the cliff, however, they merely spread their wings and are off.

Gannets are also fussy birds, frequently arranging and rearranging their nesting material as they tend their eggs or young. A gannet's single, chalky blue egg is laid in early to mid-May, in a moss-lined, bowl-shaped hollow built on top of a pile of debris. Both sexes incubate the egg, although the male seems to serve longer duty during the forty-two- to forty-four-day period. By mid-June, most of the eggs in the colony are laid and by the third week in June, the first naked chicks appear.

Gannets are attentive parents, and it is fascinating to watch them feed their young. The adult opens its beak and the youngster plunges its tiny head inside to eat the regurgitated, partially digested fish there, mostly mackerel, capelin, and herring. [29]

Insects

INSECT STRUCTURE

Most people don't like insects; bugs they call them. Butterflies are somehow exempted from this dislike; probably because they're pretty. Pretty or annoying or even frightening, we'd better learn to get along with them; half the living creatures on earth are insects. Any place you can think of has insects that have adapted themselves to that particular environment; everywhere perhaps but in the ocean deeps. They survive because they can reproduce more quickly than man can kill them. They've been around a long time too; the cockroach for instance, has been scuttling about for 300 million years or so.

Insects belong to the same class as lobsters, Arthropoda, which means invertebrate animals with an external skeleton and jointed legs. They're called insects from the word "insecta" meaning "sliced into," in reference to the three segments that make up their body.

The front segment, the head, contains eyes, antennae and mouth. Thousands of light receptors form a complex eye, while the antenna is another acutely sensitive organ picking up temperature changes and smells and used to taste and feel. Because there are so many different ways of feeding, individual species have developed specialized mouths, a chewing type for the Japanese beetle, sipping straws for moths and butterflies, while the thrip has a mouth for sucking plant juices. Mayflies, which live only for a few hours in order to mate, never eat and have no mouths at all.

The middle segment of an insect is the thorax, where the four wings are attached (although some insects have no wings) and where the six jointed legs are placed. There is wide variety in legs. The flea with its narrow body can move easily through hair, has strong rear legs with which to leap from host to host. Dung beetles have powerful front legs for digging.

The abdomen is the third segment, containing the oviduct. Insect blood is colorless and not confined to veins and arteries, but fills the entire body cavity. The heart is a long tube reaching from the head to hind part, and there often are little accessory hearts at the bases of wings, legs or antenae to pump blood to these narrow organs.

Truly, insects are amazing creatures, even those pesky mosquitos. [1]

INSECTS IN A STREAM

Insects belong to a group of animals called arthropods, whose numbers comprise 80 percent of all animals known. They have an external skeleton to which muscles are attached, which must be shed as the animal grows. An insect's body is divided into three parts: the head, thorax, and abdomen. Growing insects, depending on the species, undergo either complete metamorphosis, including the four stages of egg, larva, pupa, and adult, or incomplete metamorphosis, skipping the pupal stage and moving from egg to nymph to adult. Metamorphosis can allow insects to assume different forms and even different ways of life as they grow.

Stream insects possess amazing survival adaptations to their flowing, wet world. On a walk by a nearby stream, in the still pools at the water's edge, we may find two familiar critters flitting along the surface.

Water striders are known by almost everyone. The feathery tips of the water strider's legs allow it to skate along on the surface as it prowls for unlucky insects that have fallen in. Tiny claws set back on the legs are used to handle captured insects. These insects use the water's surface much as a spider uses its web to catch prey. Once a victim is captured, the strider uses piercing, sucking mouthparts to finish off the meal. The next time you watch some water striders, catch one of the many mosquitoes that are pestering you and toss it into the stream. Then watch the striders' feast begin.

In the bend of a lazy meander, you may encounter some other common denizens of the stream's surface—a group of frenetic, gyrating whirligig beetles. These shiny, bluish black beetles dart erratically across the water in large groups. As each day and season progresses, they move gradually farther into the open water. Whirligig beetles are the only beetles that swim on the water's surface. The soft aquatic larvae of the whirligig beetles breathe by the diffusion of oxygen through their skin. Resourceful in their survival instincts, some whirligig larvae will even pierce submerged plant tissue in quest of air to breathe. Mature larvae pupate at the water's edge.

Life as an adult whirligig beetle is fast, frenzied, and competitive. Oarlike legs keep them skittering along the water's surface in their incessant search for a tasty insect meal. Unique, two-part eyes allow them to focus above and below the water simultaneously. Occasionally a beetle will dive using a bubble of air trapped under the abdomen as an aqualung. When air in the bubble is depleted, oxygen is diffused from the water to replenish the supply.

You will never be wanting for new discoveries while sifting the gravels of a streambed.

Crane fly larvae are found in abundance in the stream that runs through our farm. Cream-colored and maggot-like in appearance, these larvae are one and a half to two inches long and bear some appendages on their head that look like the nose of a star-nosed mole. Adult crane flies resemble giant mosquitoes, but they do not bite.

Nymphs of the damselflies should not be overlooked. As adults, these are the familiar "darning needles" that children are often warned will sew their mouths shut if they don't keep quiet. In the still pools these nymphs, with three tapelike gills on their tails, are notable predators upon tadpoles and other insects. Gills along the rectal lining aid in their breathing. Water spurted out of the anal pore acts like jet propulsion. After a few years the nymphs will crawl up onto a reed, shed their nymphal skin, and emerge as a beautiful fragile adult with gossamer wings. Adults hunt by waiting in ambush until their unsuspecting prey flies by.

As you continue to turn over rocks, sift gravel, and explore the stream, a tube-shaped caddis home of sand grains or leaf pieces may happen by. Of all the insects encountered in the lucid pools and rolling ripples of the streambed, none is more curious, ingenious, and inspiring than the omnivorous larva of the caddis fly.

Caddis larvae construct their homes by weaving an intricate tube of silk threads that is closed at one end. Depending on the species, either sand grains, leaf pieces, or small sticks are glued to the silk tube, often in a neat, spiral pattern.

Caddis fly larvae are mostly herbivorous, eating moss, algae, and dead leaves. Some even build tiny, stocking-shaped nets to glean their food from the current. Abdominal gills are used for breathing, which is aided by the undulating motion of larvae in their cases to help keep fresh water circulating.

When ready to pupate, caddis will anchor their homes to a rock and close themselves in. In time, the adult, moth-like caddis flies emerge. After a few short days, during which little is eaten, mating occurs and the femal embarks on a perilous, egg-laying mission to the stream bottom.

Stream insects are true marvels of nature's design. Of ancient lineage beyond our conception, they are beings whose simple, yet infinitely successful, adaptations to stream waters have allowed them to survive through many long periods of traumatic change on earth. [2]

INSECT ANTIFREEZE

Keeping active in the cold season is important; keeping alive — and unfrozen — is essential. When living tissue freezes, expanding ice crystals destroy cell membranes, causing irreversible and fatal damage. Death comes even before freezing is complete, when there is no longer enough liquid in the cell for the enzyme activity essential to life.

Insects survive subfreezing temperatures by supercooling, lowering the freezing point of their body fluids, and either by slowing ice formation or by being able to function even when their extracellular fluids have been frozen. Both groups use antifreezes: the polyhydric alcohols, sorbitor and glycerol (chemically similar to the glycol used in automobiles), and the disaccharide trehalose.

Insects in the first group manage by reducing the chance that ice will form in their circulatory fluids. Ice tends to form rapidly around a nucleator, a tiny ice crystal or speck of dust that offers a solid frame for other molecules to attach themselves to. These insects purge their bodies of potential nucleators, principally by emptying their guts, and produce antifreezes, which lower the freezing point by raising the concentration of solutes in the body fluids (saltwater has a lower freezing point than fresh water, for example). The antifreezes have multiple hydroxyl radicals in their molecules, which tend to bond with the hydrogen in the water molecules, thus greatly reducing their tendency to aggregate into ice crystals.

The quantities of antifreeze produced are prodigious. Eastern tent caterpillars, very much alive in their winter cocoons, are as much as 35 percent glycerol by body weight in midwinter. By the end of April the glycerol has all but disappeared from their systems.

In the second survival strategy, insects tolerate some freezing of their extracellular fluids. The freezing itself releases heat (known to physicists as the latent heat of fusion), which slows the cooling of the body as the outside temperature falls. But that is not enough. These insects also use antifreezes to slow the freezing of the extracellular fluid. And the antifreezes play another vital role inside the cells. As the fluids outside the cell slowly freeze, the concentration of solutes in the remaining fluids rises. Normally this would set up a condition in which, due to osmotic pressure, the less saturated fluid inside the cell tries to flow through the membrane toward the more saturated solutions. The antifreezes, however, hold the water inside the cells, helping to slow the potentially fatal osmosis.

Whichever strategy it uses, each winter insect has a limit below which it will die. The limit can be low indeed—larvae of the parasitic wasp *Bracon cephi* can lower their supercooling point to minus fifty-two degrees Fahrenheit—but when it is exceeded, the insect freezes to death like any other organism. [3]

A BOUNTY OF BEETLES

"It is difficult to avoid the constant use of superlatives when talking about beetles," the noted entomologist Alexander Klots wrote, "for they are unquestionably the outstanding order of insects." Twenty years ago, estimates of the number of beetle species on earth ranged from 100,000 to 250,000. Klots predicted that studies of tropical fauna "will undoubtedly show that there are many more than the present estimate." Today, scientists say that one million species of beetles is not unlikely.

"Such numbers signify unparalleled success in life," Klots continued. "The beetles have shown a striking ability to penetrate to every part of the land environment where an insect can live, and to adapt to every means of exploiting the food resources available." Several families have entered the fresh waters, where they live—as adults as well as larvae—in all possible places, from the swiftest streams and waterfalls to the muddiest ponds. On land and in the water, we find beetles with nearly every type of food habit: living on and in plants, scavenging on plants and on animal remains, and preying on other insects and small animals. Relatively few are parasites, but even here some families have evolved a high degree of specialization and success.

Much of the success of beetles, Klots wrote, is due to "the transformation of the front wings into hard or tough, impermeable covers, known as elytra, that protect not only the delicate hind wings but usually the abdomen as well. With the hind wings folded beneath these elytra, a beetle can burrow, creep into cracks and crevices, or withstand hard knocks without impairing its ability to fly." Few beetles, Klot noted, are fast fliers. Compared with dragonflies and hawk moths, "they look like airborne trucks in low gear. Yet they can fly well enough for all practical purposes of mating and of distributing eggs." The majority of beetles, he acknowledged, "are relatively plain and humdrum looking." But many species have brilliant, metallic colors "that vie with those of any other animals, even of the brightest of the birds and butterflies. . . . The lure of the beautiful and the curious, together with the great abundance of beetles, has made them a favorite group with amateur collectors." [4]

BEETLEMANIA

In 1870, when the American Museum of Natural History was barely a year old, it received a gift of 3,800 beetles from Baron Osten-Sacken, the Russian consul general in New York. Since the museum was only an incorporated idea at that time, the Baron's beetles were temporarily housed at Brown Brothers, a prestigious Wall Street banking firm. These beetles were the modest beginning of the museum's now formidable collection of insects and spiders.

Today the beetles and other insects are housed in row upon row of gleaming white cabinets on the fifth floor of the museum, just one flight up from the dinosaurs. The faint odor of mothballs in the air reminds one of grandmother's attic. At last count there were 15,901,687 insects and spiders inhabiting about 90,000 square feet of storage space. For those who like statistics, we can break this figure down to (approximately):

    1,100,000 spiders (the largest collection in the world)
    1,600,000 beetles
    800,000 flies, mosquitoes, gnats, and others of that ilk
    400,000 bedbugs, lice, aphids, and other species having piercing and sucking mouthparts.
    8,200,000 social insects—wasps, ants, termites, and bees (including 5,500,000 gall wasps collected by Alfred Kinsey before he turned to sex research)
    2,000,000 butterflies
    1,700,000 other kinds of bugs

The most important specimens in the collection are the 18,440 definitive types. These are the actual specimens that scientists use to describe and name a new species. If at any time in the future—even centuries hence—scientists have a question or disagreement about the identity or description of a species, they must examine the type specimen, the legal representative of its kind.

Insects and spiders make up fully 45 percent of the museum's entire collection of 35 million specimens and artifacts. [5]

HONEYBEES

The honeybee has a very sophisticated social system. The females, queen and workers, are the only industrious members, with the male's sole function being to fertilize the queen.

The larvae of all of the bees begin life identically, all fed royal jelly for the first three days. The queen bee is treated differently, being fed royal jelly for the entire sixteen days until she emerges. The worker, a sterile female, and the male drone are fed bee bread (pollen and nectar) and honey as larvae; then the larvae spin a cocoon, pupating for twelve days for workers and fifteen days for drones.

Once hatched, life differs drastically for the three types of bees. The queen first searches out and drives off other queens, also destroying any queen cells to establish that she is queen. She moves around the comb for a week and then ventures out for a mating flight. A few days later she starts laying eggs, producing in theory up to 3,000 eggs per day, but in reality less. This arduous pace is kept up from early spring to late autumn for up to five years. The queen controls the sex of the bees by fertilizing the eggs that will develop into female workers and laying unfertilized eggs that will develop into male drones. As we have said, the male drone has only one purpose: to fertilize the queen, which need only be done once. The males are driven out in the fall leaving only the females to overwinter.

The worker bees are truly the backbone of the hive. During their life cycle of thirty-five to forty-five days, they perform the following functions in chronological order from their emergence: cleaning cells and keeping the brood warm (two days), feeding older larvae (three days), feeding younger larvae (six days), producing wax, building combs, transporting food within the hive (six days), guarding the hive entrance (four days), visiting flowers, pollinating them, and collecting pollen, nectar, propolis, and water (remaining days). As eggs are laid constantly by the queen, all functions are performed simultaneously by different-aged workers. The bee born in the fall is not able to forage and lives through winter to start the new brood cycle in spring.

So, you have to be very active to claim to be as busy as a bee! [6]

BUMBLEBEE

While working in our flower garden last week we somehow disturbed a hardworking bumblebee, who flew off with an angry buzz from a columbine blossom. The honeybee, also hard at work in the petunias nearby, is really an import brought to this country by honey makers, while the bumblebee, sometimes called a "humblebee" by old-timers, is America's only social native bee. Social means that the bees live in colonies with a queen and her offspring. Some of the daughters do not mate, and it is these infertile females that do most of the work in the colony. The bumblebee is different from his imported cousin, the honeybee, in that the colony does not continue year after year. In late summer all the bees die except for a young fertile female, the queen, who crawls into a safe place for the winter. In spring she emerges to give birth to a new colony, often in a deserted mousehole or a chipmunk nest. Here she makes a honey pot and lays her eggs in waxen cells that she has stocked with pollen. The larvae hatch, eat their food, supplemented by food given by the mother, and when fully grown spin cocoons in which to pupate. During the pupa stage the larvae are immobile and eat nothing. Three weeks after the egg is laid, the grown-up bumblebee crawls out of its cocoon ready to start its life work: gathering food for its future brothers and sisters. In the late summer a young queen will mate to carry on next summer a new colony of bumblebees. [7]

BEES AS UNDERTAKERS

At different stages of their lives, worker honeybees may clean out cells, feed larvae, process pollen, guard the nest, and forage. Now, Cornell University entomologist Kirk Visscher has found that a few of them also perform another job—undertaking.

Entomologists have long noted that honeybees remove corpses from the hive, but no one knew that undertaking was a specialized job, limited to a few individuals. The job is an important one; if it is ignored, dead bees would accumulate at the rate of more than a quart a month, carrying disease and attracting vermin.

While watching bees in glass-walled colonies, Visscher noticed that most workers that encounter a dead bee ignore it, lick it briefly, or tug at it momentarily. Some bees, however, purposefully approach the corpse, grasp the lifeless bee in their mandibles, drag it through the hive, and fly up to four hundred feet away before dropping the body. By painting dots on bees that carried away the dead, Visscher estimated that one to two percent of the hive act as undertakers. The job is only temporary, however, lasting a few days before the bees go on to another job.

To ascertain whether undertaking is a vocation separate from a bee's regular cleaning chores, Visscher put balsa wood models of dead bees and freshly killed bees into the hive. The bees carted out the corpses in about seven minutes, but it took seven hours before they got around to the wooden models.

Visscher suspects that bees, like ants, may emit a chemical signal when they die, prompting the undertakers to their duty. He has not isolated such a chemical, but he has indirect evidence that one exists. He placed dead bees coated with paraffin, corpses purified by solvents, and freshly killed bees in the hive. In about three minutes, undertaker bees removed all the fresh corpses. It took them about three times as long to clear out the waxed and treated bodies. [8]

BEE BITES

When a honeybee stings you, she bends her abdomen sharply downward and extrudes the sting that is ordinarily concealed in a sting chamber. With a quick jab, she sticks the tip of the sting into you. The sting has two lancets that have barbed tips and are capable of being moved independently. Muscles attached to the sting contract in rapid alternation advancing the lancets, the barbs on one securing it while the other is pushed in deeper. Meanwhile, the same muscles that are embedding the sting are also pumping venom down the poison canal where it escapes through a cleft near the tip of the sting shaft.

The most remarkable thing about the honeybee sting is that it's like an automatic hypodermic syringe. Even after the bee has flown away, the muscles attached to the sting keep working it in deeper and deeper and keep pumping in venom. If you remove the sting from yourself in a few seconds, you'll cut way down on the amount of venom that's injected. It's no good to grab the sting between your fingers, because you'll just squeeze the rest of the venom into you, just like squeezing the top of an eyedropper. Instead, take your fingernail and scrape the sting out. If your fingernails are too short, use a penknife blade or a nail file. You'll find that the sting comes out easily.

From the bee's point of view, having a sting that tears out, even if it is ultimately fatal to the bee, is a neat way to guarantee that more venom gets into the victim. If she had to hang around, she might be brushed off prematurely. In general, of course, bees don't sting people. You have to be disturbing the hive to get attacked, or else step on or bump into an individual bee somewhere. Even if you throw rocks at a beehive, less than half of one percent of the bees in the hive are likely to sting. Alarmed bees look for movement and color, and at close range, odor. Dark clothing, hair, and leather elicit stinging. Away from the hive, brightly colored clothing and sweet perfumes may attract a bee, and chemicals in a person's breath may cause her to sting. [9]

LIGHTNING BUGS

The first summer evening's walk, when the fireflies are glowing their signals like time lapse photography across the darkness, is always a special treat for me. The unified rhythm of their flashes is almost mesmerizing.

Then, as if to bring me back to earth, there will be intermittent lighting signals from nearby on the ground. It is easy enough to pick up one of these lightning bugs. Their flashing usually continues, and you can watch as special abdominal segments light up, on and off, with their cold yellow-green light.

I've often wondered why some fireflies light up during flight and others emit their light signals from the ground, but only recently did I learn the answer.

Those seemingly lazy ground-hugging fireflies are the females, many of which are wingless. They respond to the light signals of the males flying overhead with a similar flashing pattern from the ground. If he's alert, the male drops to the ground beside her; if he's lucky, they mate; if he happens to have been lured to the ground by mimic signals of a female of another species of firefly, he may well be eaten.

Fireflies belong to the *Lampyrid* beetle family of which there are many species. To ensure maintaining species identity, the signals of fireflies are species-specific; each species has its own unique flashing pattern. Duration and number of light flashes differentiate one from the other, and only those of the same species respond to each other (except for the mimic predators).

The intense light emitted from fireflies, and even from their egg and larval (glowworm) stages, has intrigued scientists because less than one percent of the energy is wasted as heat, where electric lights waste between seventy-five and ninety percent. Simply stated, fireflies oxidize a fatty substance called luciferin in the presence of an enzyme called luciferase. The scientific process is understood, but not how to emulate it in an economical way.

Notice the different lighting signals of male fireflies this summer and look to see whether similar light responses glow from the ground. [10]

GRASSHOPPERS

Walk through a field on a sunny September day, and the grass comes alive with fleeing grasshoppers. Motionless, they are hard to see; in motion they are hard to miss as they pop out of the depths of the grass and leap to shelter many feet away. To spot, follow, and catch a grasshopper is a challenge.

Once you catch one, close examination of a grasshopper reveals a creature that is far more interesting in color, design, and texture than the dull brown flying capsule one glimpses leapfrogging from blade to blade. The undersides of our brown grasshoppers are pale yellow, and delicate yellow stripes outline the wings and the muscular thighs of the hind legs. The herringbone pattern on these legs comes from efficiently designed scales that give flexibility and strength. Under a hand lens, the bristles and claws projecting from the legs look sharp and prickly, like weapons. In fact, they assure a good grip, no matter where the grasshopper lands.

The fable of the grasshopper and the ant that depicts the grasshopper as a frivolous fiddler who does not survive winter because he does not bother to prepare for winter's hardships, and the ant as a hard worker who earns survival through the cold months, is a little unfair to the grasshopper.

Insects have many strategies for surviving the winter. Social insects like ants and honeybees join in a communal effort to provide shelter, food, and warmth to get through the cold months. Solitary insects, like the grasshoppers, survive the winter in the stage of their life cycle best adapted to withstand frigid temperatures: the egg stage.

The frantic activity in the sun-warmed fields marks the final maturing stages for this year's grasshoppers. Early in the summer, eggs laid last fall hatched into tiny grasshopper nymphs, miniature replicas of adults. Gradually they grow and mature, each stage marked by the shedding of the exoskeleton. For the red-legged grasshopper it takes fifty days from hatching to adulthood. The nymph doubles in size, and after the final molt has fully formed wings.

Before the weather turns too cold, the mature grasshoppers must mate and the females deposit their eggs in the ground. Then, as the fable suggests, the adults die — not because they were lazy, but because for grasshoppers, the protectively coated eggs snug in the earth can best survive the winter. [11]

CRICKETS

An early autumn evening would not be complete without the chirping of resident crickets. As the nights grow longer and cooler the music of these nocturnal creatures increases in intensity. They have the deadline of hard frosts to work against, for several successive nights of subfreezing temperatures will bring their lives to an end. Before this occurs, mating (of which singing is a part) and egg laying take place.

The male black field cricket is responsible for much of the music we hear. With the edge of one wing rubbing against the opposite wing, he creates the chirping noise that is his means of claiming territory for himself and his mate. In the presence of a female field cricket, the series of chirps becomes much higher in pitch, and the male cricket moves rapidly about. Some of these sounds are made at the rate of 17,000 vibrations per second — hardly distinguishable to humans.

In many countries, especially China and Japan, crickets have been so prized for their singing that they are kept in cages and carefully cared for. Actually, since A.D. 960 in China crickets have been kept for their fighting ability. Cricket fights were once as popular as horse racing is today. (The Chinese fed the crickets special diets, including mosquitoes fed on trainers' arms, and weighed the crickets in order to classify them.)

The next time you look closely at a cricket, notice whether or not it has a long, spearlike appendage protruding from its abdomen. If it does, it's a female, for this is an egg-laying device, the ovipositor. If you hear a cricket singing, you immediately can assume it's a male, as females make no noise.

Crickets are considered by many people today as symbols of good luck, and they are a welcome addition to household and hearth. They do add cheer to a silent home, but beware — they are voracious vegetarians, devouring everything from clothing to books. [12]

BUTTERFLIES OF SPRING

In the month of June, many people begin to notice and remark about the large numbers of butterflies in gardens and fields.

Most of the butterflies of spring emerge as adults after spending the winter as a chrysalis. They probably possess some kind of internal clock and temperature sensing device that tells them when it is safe to emerge. Butterflies need fairly warm temperatures and nectar-producing flowers to feed upon. By late May, trees and other plants are well-leaved out, ready to feed the new brood of caterpillars that will emerge from eggs laid by the adults.

A few butterflies do not spend the winter in the chrysalis stage but as adults. Mourning cloaks, brown butterflies with yellow borders on their wings, hibernate in crevices and are lured out by the very first warm days of spring. The closely related tortoise shell butterflies have similar habits, but these brown, yellow, and orange mottled insects are rarely-seen mountain residents.

The tiny common blue, or spring azure, is one of our earliest and most common butterflies, but because of its one-inch size, it often goes unnoticed. As both names indicate, the spring azure is a blue butterfly, which was described as a "violet afloat" by Sam Scudder, a pioneer in the study of American butterflies. Spring azures have a trait that was not recognized by early naturalists: the first butterflies of the season are bright blue, but the one or two broods produced during the summer are progressively paler, with lighter and more diffuse spots on the undersides of the wings. As the butterflies seen later in the summer were a different color from the familiar spring azure, early naturalists did not recognize them as the same species.

Some of the most abundant butterflies are the medium-sized sulphurs and whites, but the largest and most showy butterflies are the swallowtails, named for the tail-like extension on the edge of the hind wing. The black and tiger swallowtails are the most common species. Black swallowtails have some yellow and blue spots on their black wings. Tiger swallowtails are yellow, with black stripes across the forewing, and black, blue, and orange markings on the hind wing. They are the large butterflies that are often seen gathered around puddles on dirt roads, following a habit that provides food for eaters of dead butterflies.   [13]

THE MONARCH

The monarch butterfly is large and orange and black — and possibly slightly poisonous.

One must say "possibly," for not all monarch butterflies contain the mild dose of heart poison that makes any bird that bites into it vomit.

The folklore of butterflies has all monarch butterflies as bitter-tasting insects that contain sufficient chemicals to discourage birds from ever biting another monarch. It is not true. The butterflies as caterpillars accumulate the chemical from eating wild milkweed leaves, which contain the heart poison.

It so happens that the common milkweed of New England does not contain the chemical. The common milkweed was near enough to a flower for Governor John Winthrop to send samples to England in 1670. It proved so popular that it was planted all over Europe and into the Middle East. In fact, a century later, when the great Linnaeus was dealing out scientific names for plants, he mistook the common milkweed as originating in the Orient rather than in the New World and gave it the species name of *syriaca*. Maybe we should value our milkweeds more.

Monarchs feed upon milkweed and nothing else. There are twenty-five species of milkweeds in New England, all presumably supporting a monarch population, and some of those milkweeds must contain the emetic chemical, because it occurs in monarchs migrating south over New England.

The person who has done the most in calling attention to the variable nature of monarch palatability is Lincoln Brower, biology professor at Amherst College. Among monarch butterflies passing through western Massachusetts, Dr. Brower has found one in four contains a dose of heart poison that can cause a blue jay to vomit. The bird eats the butterfly, becomes ill within twelve minutes, then spends thirty minutes recovering. Brower feeds butterflies to blue jays under laboratory conditions. Repeated experiments indicate that the blue jay comes to associate the illness with the butterfly.

The monarch has a look-alike called the viceroy, a butterfly that never eats milkweed and never is poisonous. For many years, the viceroy has been considered a mimic that benefits from the poisonous effects that the monarch has upon birds. Experiments indicate that a blue jay that has eaten a monarch that contained poison will reject the viceroy.

For many years this association between the monarch and the viceroy has been recounted by naturalists as one of the tricks that nature plays. Brower's investigations over the years indicate that the deceit is much deeper than most naturalists supposed. Now it seems clear not only that viceroy butterflies benefit from mimicking monarchs but also that other monarchs, many of whom are quite edible, benefit from displaying the monarch color pattern. [14]

PILL BUG

As the saying goes, wood heats many times over; when it's cut down, sawed up, split, hauled, piled, and finally burned. Most of you have probably shared in at least some of the steps, and if so, you may have noticed lots of various little creatures scurrying for shelter as you moved the logs.

Small, hard-coated, many-legged creatures scuttling through drying wood do not appeal to many people, but watching them can provide a welcome diversion to the bend-down-pick-up routine of wood stacking. The one I see most often looks to me like a miniature armadillo, especially if it goes into its rolled-up don't-touch-me act. It is called a sow bug, or pill bug.

Sow bugs are arthropods. As such they have hard outer

body coverings (exoskeletons) and segmented bodies. Unlike most of their crustacean relatives, lobsters and crayfish, for example, sow bugs (*Isopods*) have no outer layer of platelike shell on their backs. Their bodies are flattened, and they have little leglike projections but not typical legs.

There are good reasons why pill bugs are found at the bottom of log piles. Food is plentiful there for these scavengers, who eat decaying animal and vegetable matter, each other, and their own shed skin following a molt. Also they need moist surroundings in order to breathe. They breathe through gills similar to their lobster and crayfish cousins, so the air surrounding the pill bug must be moist enough for the blood running through the gills to absorb oxygen.

Mother pill bugs have something in common with kangaroos and opossums; their young are born, then remain in the mother's brood pouch for awhile. From birth, the little ones look like their parents; they have seven thorax segments each with a pair of legs, and six abdominal segments from which their gills extend on what look at first like more legs.

The next time you need a break from stacking wood, look around for pill bugs, and think about some of their remarkable adaptations for living under logs.   [15]

APHIDS

While weeding in the vegetable garden recently, I came across several plants whose stems were covered with small, dark bugs. Little did I know that my vegetables were serving as hosts for some amazing, although somewhat harmful, visitors. These innocent-looking insects are called aphids, and they are the bane of many a farmer's existence.

Aphids are colonial, in that many of them inhabit one

small area, such as a plant stem. Aphids live off plant juices and may be seen attached to their host plants in a variety of positions, but always with their tubelike mouthparts inserted into the plant stem as they suck its juices.

If you look closely at a cluster of aphids, you will find several different forms. The most abundant form at midsummer is one that lacks wings. These are females, which give birth to living young and do not lay eggs. They do this until the plant is overcrowded with aphids and the food supply seems to be dwindling; then another form of aphid, which has four wings, is produced. These fly away to another plant and start a colony there. When the weather begins to get cool, male and female aphids are developed, and the female lays eggs that overwinter and hatch in the spring.

While aphids take in plant juices, they secrete a sweetish liquid called honeydew, which is greatly sought after by bees, wasps, and ants. Bees and wasps will come along and rob the feeding aphids of this secretion. Some species of ants, however, actually care for the aphids, stroking them with their antennae until the aphids respond by producing a drop of honeydew, which the ants then consume.

One specific aphid, the corn-root aphid, actually has become largely dependent on one species of ant, the cornfield ant. The ants store the aphid eggs in their underground nests and care for them throughout the winter; in the spring, after the eggs have hatched, the ants transfer them to the roots of various plants on which they feed. Later, when corn plants have become available, ants move the aphids to the roots of the corn.

Although aphids appear defenseless, occasionally they are able to defend themselves against potential predators such as ladybugs. Through two small tubes at the end of their body they secrete a waxy substance that they smear in the eyes of the attacking insect, thereby creating enough of a distraction often to allow escape.

When you're next out weeding, and you spy a colony of resident aphids, before you do away with them take a moment to examine them under a magnifying glass. Your curiosity will be rewarded with a fascinating sight. [16]

CRAB SPIDERS

Late summer and the early fall, before the first hard frost, is the time when spiders tend to be noticed most frequently. These shy creatures are often feared and maligned unfairly. A spider will make every effort to get out of the way of a human, but, if necessary, will resort to its only means of self-defense, its bite.

Although spiders are like insects in having jointed legs and hard external skeletons, they really are very different. Insects, in the class *Insecta*, have six legs, a body divided into three parts, one pair of antennae, and wings (usually). In contrast, spiders, in the class *Arachnida*, have eight legs, a body divided into two parts, and no antennae or wings. Spiders usually also have eight simple eyes, chelicerale jaws that are tipped by fangs connected to a gland containing a paralyzing and digestive fluid, and pedipalps between the first legs and the jaw. In male spiders these pedipalps look like miniature boxing gloves.

Undoubtedly the skill for which spiders are most well-known is the spinning of silk. Through spinnerets at the back of the body, all spiders spin silk. At least seven separate types of silk can be spun individually or in combination with others, creating an unbelievable variety of threads. Different types of silk are used for lining retreats, weaving egg sacs, wrapping captured prey, and creating the continuous dragline left behind by all spiders as a safety line or escape route. Another amazing use of silk is for travel by ballooning. Indian summer is a favorite time for ballooning, when spiders climb high and release multiple lines of silk until the wind picks them up and carries the animals off. Spiders have been found as high as ten thousand feet up and as far out to sea as two hundred miles using this method of travel.

Probably we are most familiar with the spider using its silk in forming a trap for its food. There are very few animals that actually trap their food — humans, some caddis flies, fungi, gnats, and spiders. But many people are unaware that there are spiders that do not use webs to catch their food. One of the most beautiful yet rarely noticed of these is the crab spider.

There are approximately two hundred species of crab spider in North America. They are known as crab spiders because their long, curved legs are held out to the sides, and because they frequently move sideways, looking quite crablike. Except for the dragline, they do not spin silk, but take advantage of their excellent camouflage to hide and ambush their prey. They are found frequently in leaf litter and low vegetation, as well as on flowers, blending in with the background color. These flower spiders vary in color from a bright pink to the more common whites and yellows. A yellow spider hidden away on a goldenrod or mullein is next to impossible to see. If transferred to a flower of a different color, the spider will either return to a flower on which it is camouflaged or will gradually make some change in its color to match its new environment. With its excellent vision and camouflage, the crab spider has evolved a fascinating and successful means of catching its food, taking advantage of the flower's allure to insects. [17]

BLACK WIDOW SPIDER'S BITE

Even its name sounds dangerous—black widow spider. But it's a good name for the dominating female of this poisonous spider species. Unless he is very careful in his courting procedures, the male spider may be eaten alive by an angry female black widow. The female black widow should be avoided by humans, too. While her bite is seldom fatal to people, it can be very painful, producing symptoms not unlike those of appendicitis.

Luckily, the female black widow would much rather avoid people than bite them. Even when disturbed in her nest, usually she will try to escape rather than attack. And she's also easy to recognize. The female, which can grow up to two inches in length (twice the size of the male), has a tiny yellow, or sometimes red, patch, often in the shape of an hourglass, on her abdomen. Her abdomen can swell to half an inch in diameter when it's full of eggs. Found throughout the United States and other parts of the world, but mostly in warm climates, the female black widow spins her sticky, untidy web in dark, dry places — under stones, in holes in the ground, around tree stumps, in log piles, garages, and basements.

In the United States and Canada, fatalities from wasp and bee stings far outnumber those from spider bites. While her venom sacs are small, the female's poison is up to fifteen times more potent than that of a rattlesnake. A black widow bite often causes nausea, swelling, and mild paralysis of the diaphragm, but most bite victims recover without serious complications.

The male black widow's poison is weak and ineffectual, so no wonder he's such a careful suitor. He approaches the female's web and taps out a kind of Morse code on its threads to find out if she's ready for him. If she's not, she may turn on him in a fit of black widow anger. Or she may mate with him and eat him later.

More often, the female black widow, like most spiders, feasts on insects. When a victim gets caught in her web, she wraps it tightly in silk then swiftly bites it with two tiny fangs that deliver her potent poison.

Male black widows don't feed. In fact, they're very seldom seen at all. They spend most of their time wandering in search of females — which for them is a hazardous occupation. [18]

Water & Aquatic Life

WATER

A place water goes where it shouldn't is simply to waste. For example, 40 percent of the water in the modern American household is used for flushing toilets. Kennedy P. Maize points out in "The Case for the Waterless Water Closet," in the July 29, 1978, issue of *Environmental Action*: "One person using a flush toilet for a year pollutes 13,000 gallons of fresh water to remove 165 gallons of body waste. We take clean water and mix it with potentially fine fertilizer, making both useless. Then we spend lots of our good money trying to separate them again."

Composting toilets may be the way to go in the future. Those with no choice at the present except the standard toilet can reduce the volume of water used per flush by a plastic bottle, filled with water and weighted with pebbles, to displace water in the tank.

About thirty percent of the water used in the home goes for baths and showers. Simple devices can reduce the use of water by half or more, and not so incidentally they also reduce the use of energy to heat all that water.

Using the washing machine or dishwasher only for full loads conserves water. Quick repair of pipes and appliances keeps water from leaking uselessly. And, of course, there's the chorus from everyone's childhood, true as ever: "Turn off the faucet when you're not using the water."

In addition to leaks and waste, there is also the problem of water that is simply unaccounted for. A system that can identify where 80 percent of the water that goes into the system comes out again is considered by water supply experts to be doing reasonably well.

When Frontinus did his work on the Roman water supply, metering was simple. The user of water was given imperial permission to take water out through a hole of a certain diameter; the metering device was simply a piece of lead pipe of a particular diameter, stamped with the imperial seal. Water could flow out of the hole continually. Obviously it was easy to have metering problems, accidentally or intentionally.

While metering is more sophisticated now, it remains a problem. Meters must be maintained continually to be effective, and the price of water must reflect the actual cost. Cheap water encourages waste. [1]

WATER TABLES

The underground water table gets its supply from only one source: the moisture that falls on the surface of the land in rain or melted snows. If the water from the rains or melted snow runs off the surface of the land too fast, it does not have a chance to soak into the ground. Anything that speeds the runoff, therefore, robs the underground water table of its normal supply. This moisture must make its way slowly down into the soil until it comes to a stratum of rock or impervious clay where it can go no farther. There it is held in storage for the many uses that nature requires.

As moisture thus absorbed increases in quantity, the level of the water table rises just as the surface of the water rises in a tub when more water is added. If it falls, it is a sign that the new supply of moisture has for some reason been prevented from working its way down through the soil. It has been thought that a falling water table is caused by a decrease in general rainfall, which of course is a very simple explanation. But examination of records over a period of several wet years has shown that the water table continues to fall even in periods of heavy rainfall, and only a little less rapidly than in a dry cycle of years.

There was a time not many years ago when rains and melting snows were held on the surface of the land in marshes, ponds, and shallow lakes. In those pools the water stayed often year-round, giving it plenty of time to soak into the soil and replenish the water table. Millions of acres of these marshes and sloughs, and even some of the shallow lakes, have been drained off through man-made drainage ditches in order to make more dry land for farming. That is one thing we know lowers the water table.

Forests and underbrush, with their thick carpet of old leaves and decaying logs and deep matted roots, once occupied much more of the land than now, and rain and snow falling in the great natural forests was held in the spongy blanket of vegetation until the moisture slowly seeped down into the ground to join the underground water supply. When we clear the forests and underbrush from the land, we destroy another of nature's methods of retarding the runoff of surface waters. The water is gone before it has time to soak into the ground.

Our prairies and meadows, when the pioneers first saw them, were waist-high with a heavy growth of native grasses, which caught the snows and rains and held the water in their matted roots almost as effectively as the forests. When we plow them for planting, or when our sheep and cattle graze them down close to the ground, they no longer hold back the moisture until it has time to seep into the earth.

Now, as though lowering the ground water table is not enough of a calamity in itself, man is guilty of a double crime. By the same acts with which he destroyed his own habitat, he has robbed wildlife—songbirds, fish, wild ducks and geese, and fur-bearing animals—of their natural homes. Their breeding grounds have been destroyed and their food and water supplies just as badly affected as man's living conditions. No matter how carefully we protect wildlife from human molestation, they cannot multiply when their natural homes are destroyed. [2]

WOODLOTS AND WATER

The most important thing that woodlots do for us is not to grow the lumber that builds our homes; it is not to grow the fuel that heats our homes. The most important thing woodlots do for us is to provide pure water, and provide it in benign amounts. Water and woodlots are inextricably interrelated. Without water woodlots couldn't be; without woodlots—or forests—flood and drought would make much of the earth uninhabitable.

Water is nature's most powerful tool. With gravity and water, all mountains are destined to become plains. You will have seen in your own experience the awesome power of flood. But probably few will have noticed how small are the beginnings of flood. A single drop of water falls from a height. Photographs show that the single drop makes a tiny depression in unprotected ground. As the water splashes out of this depression, it carries minute amounts of soil with it. On any slope (and, by the nature of things, there is no such thing as absolutely level land) this single drop with its soil burden rolls downhill. It is joined by billions of similar drops all carrying bits of dirt. The force of this moving mass picks up more and more load and soon the drops start to coalesce into streams, brooks, and rivers—flood waters, bent and determined to level the land, by their own sheer force, and by the abrasive action of their soil load.

Whether by happenstance or divine plan, forests slow the gathering of floodwaters to a turtle's pace. Forests, including small woodlots, protect the soil against the impact of falling water, the small beginning of flood. Much water that falls in a forest never reaches the ground. Rain and snow lodge among leaves and branches and are quite quickly evaporated back into the air. The water that does fall to the ground in a woodlot is held; it doesn't run off—or at least it doesn't run off as quickly as it would from unforested ground.

Run-off from bare soil is almost total. Grass slows the rush of water. Forest soil, however, holds water and controls flood better than anybody's Corps of Engineers. How is this miracle accomplished? By the nature of forest soils. Many characteristics of forest soil contribute to its water-holding properties. [3]

TRENDS IN NEUROBIOLOGY

A freshwater lake in New England is closed to swimming after a large fish kill. The taking of shellfish along the Maine coast is prohibited during an invasion of red tide. Scientists flock to marine laboratories worldwide to study the giant nerve fiber of the squid. Neurotoxins, acting on sodium channels in nerve fibers, cause the fish kill and red tide toxicity. Scientists study squid nerves because the giant nerve cells of certain marine organisms are useful in the study of neuronal function. Investigators may be able to solve the mystery of Alzheimer's disease, which produces premature senility in humans, by studying the nerve cells, or neurons, of squid, lobsters, and marine worms. And there is hope that this research eventually will provide answers concerning many other nerve and muscle disorders.

Blue-green algae undergo periods of excessive growth in some lakes. One such species, *Aphanizomenon flos-aqua*, produces one of the most poisonous substances known, a toxin that directly attacks the basic mechanism by which the neurons of higher organisms produce the impulses that carry on the function of nervous systems. It is now known that this toxin is either similar or identical to the toxin produced by the red-tide organism in the ocean. Humans swimming in lakes rich in this toxin are in the same jeopardy as people eating shellfish contaminated with red-tide toxin. Despite these dangers, this toxin has been most useful in revealing the mechanism by which nerve fibers conduct information. In this regard, the humble squid has been a favorite species for study.

The giant nerve fiber, or axon, of the squid has played a key role in helping scientists understand the workings of all nerve cells. Using squid giant axons, they have studied the mechanisms whereby nerve fibers generate electrical impulses and use them to send information along biological transmission lines. Several recent developments in nerve cell research bring us ever closer to understanding the molecular basis for electrical events in nerve cells.

Another area of nerve research is the study of how a neuron internally transports materials important to its function and maintenance. This is one of the most amazing mechanisms known in living organisms. As in other nerve research, the squid, along with lobsters, horseshoe crabs, and three different marine worms, has become an important tool because of the large size of its axons. These creatures have figured prominently in unfolding the wonder of the structural basis for neuroplasmic transport.   [4]

LEAKING GAS TANKS

Leaking underground petroleum storage systems at service stations and corner grocery stores represent the most serious groundwater pollution in Maine and the nation today.

How serious is the problem? Tank testing surveys in various parts of the country indicate that twenty to fifty percent of underground storage systems leak. If we assume that twenty-five percent of the tank systems in Maine leak at a rate of half a gallon per hour, we can calculate that 30,000 gallons of petroleum leak into the ground every day, or 11,000,000 gallons are lost every year in Maine alone. Each gallon of gasoline leaked into the ground has the potential to contaminate three-fourths of a million gallons of groundwater to a concentration of one part per million, making it undrinkable. In the past two years, the Department of Human Services has confirmed 106 cases of petroleum contamination in wells. More drinking water wells in Maine are contaminated with petroleum than any other toxic substance.

Burying things in the ground has always seemed like a safe thing to do. Indeed, underground storage has much to do with the fact that gas station explosions are few and far between.

The problem is, while nearly everyone realizes that automobiles are made of steel and consequently rust and wear out, few people seem to realize that steel tanks buried in the damp ground also rust. The average life expectancy of a underground tank is fifteen years. Because the 1960s were a boom period in gasoline consumption, and consequently, gas tank installation, a great many of our underground systems are at the end of their useful lives. A recent survey of oil distributors indicated that nearly one-third of all tanks in the ground are sixteen years old or older. Another oil industry survey revealed that forty percent of tanks dug up because of leaks had more than five holes in them. Since holes are not likely to occur simultaneously, but probably months or even years apart, the conclusion is that many tanks leak for a long time before they are removed. This is partly because the large volumes of gasoline which pass through underground tanks make even a sizeable leak appear as only a small percentage of the total volume pumped. For example, a hundred gallon-a-month loss at a station which pumps 30,000 gallons a month represents a loss of only three-tenths of one percent. In addition, many retailers do not keep the kind of detailed inventory records needed to detect leaks. Thus many underground leaks are not discovered until petroleum appears in neighboring wells, sewers or basements. [5]

WETLANDS

Man has always had a rather ambivalent view of wetlands. On the one hand, he has exploited them wherever possible for their resources. For example, peat was dug from bogs for fuel and medicinal uses and is still used for horticultural purposes, and in the past the open meadows were valued for their forage grasses — although marsh haying was much more common on coastal wetlands. More importantly from an historical point of view, the beavers that were trapped in the wetlands of North America played a key role in the exploration and foundation of this country.

But there is also an irrational aspect to man's view of wetlands. Bogs and swamps traditionally have been the dwelling places of spirits and demons. Starting with the epic *Beowulf*, in which the monster Grendel emerges from the fens to ravage the surrounding countryside, English literature and folk tales have made constant allusion to the ominous nature of wetlands. Even the language reflects this attitude. The word *heathen*, for example, was used to describe those brutish non-Christian people who dwelt on the heaths and moors.

Considering this background, the primal fears that struck those first English colonists who met with that dark wall of the wooded swamps of North America can only be imagined. It is not surprising that they took it upon themselves to drain the wetlands and reshape the face of the East Coast.

As a result of this cultural heritage, it has been estimated that 50 percent of the original wetlands of New England have been destroyed. In Massachusetts, those that remain make up approximately 6 percent of the total land area. And while this may not seem a significant amount, it should be borne in mind that these wetlands are the last vestiges of open space in many communities. In spite of this, and maybe even because of it, a large number now are under pressure from development. And this comes at a time when we are just beginning to understand the importance of their role in many of the natural systems that sustain man. [6]

SWAMP BUBBLES

Bubbles break loose from the peat-black muck and float lazily to the pond's surface. Tiny ripples of ringlet waves roll over the water's still reflective face. Walking along the shore, a child cries out, "Look at the bubbles; there must be a turtle or a frog down there."

Are those bubbles really made by aquatic critters moving along the bottom? Why would a frog or turtle want to blow bubbles while trying to hold its breath underwater? The answer lies in the soft sediments of the pond.

Pond muck is wet soil. The plant and animal remains found there must decay as they do on land. But because pond soil is saturated and air is almost absent, it takes highly specialized life forms to accomplish the vital task of decomposition.

The bacteria and fungi of the pond break down organic remains into smaller elements that are nutrients for other pond life. Bacteria live on chemical energy that is stored in the dead plant and animal matter. Gases are given off as this digestion occurs, such as ammonia, hydrogen sulfide, and methane. Hydrogen sulfide creates the rotten egg smell of swamp gas, and methane is highly flammable. One researcher calculated that a shallow, productive pond can give off ninety cubic feet per acre per day of highly combustible swamp gas.

Catch these rising bubbles and you can make a swamp torch. Take a one-quart, wide-mouthed juice bottle and submerge it in the pond, letting it fill with water. While still holding the jar underwater, turn it upside-down and catch the gas bubbles that rise as you probe the mud with your shoe or a stick. Light the gas and it will burn blue for about thirty seconds.

Don't use a larger container. A young boy once tried this experiment using a washtub and so created a small bomb.

When bubbles break the surface of a pond there may indeed be a critter lurking below, but it is probably just knocking loose a swamp gas pocket. Perhaps the mysteries of the murky pond waters can best be remembered with this "Swamp Gas Ballad:"

> Who knows a feeling that can beat
> The feel of mud beneath our feet?
> It slips and slides between the toes
> That walk between the cattail rows.
> The bubbles that come floating up
> Live in the gushing, oozing glup.
> And if they meet a lighted match
> Beware the blue, hot swamp gas flash.   [7]

PEAT

Peat is formed when plants, usually mosses and sedges, partially decompose in water. The main constituent of bogs, it is used for fuel and other purposes. Peat sods are lightweight, rectangular chunks of fibrous, brown vegetation, dried and cut specifically for fuel.

Overall, there are no accurate figures for the total of world peat bogs, but they make up less than 3 percent of the earth's land. Even so, that is a significant area. In the United States, it is the biggest energy reserve other than coal. Minnesota alone has about seven million acres.

In Ireland, peat is more than a reserve. It's the primary fuel for people like the Rosneys. Dennis Rosney, sixty-seven and now retired, has been working his perch—the area in which he has the rights to cut turf—since 1946. He wears the traditional clothes of an Irish farmer, tweed hat and jacket, and the indispensable Wellington rubber boots.

A few hundred yards into the bog, the ground quivers with each step. Only the fibrous upper crust keeps you from sinking. An undrained bog is 96 percent water, and it is no place to walk if you don't know the area. Where Dennis is working, channels have been cut to the nearest stream, draining the bog naturally, a process that can take up to five or six years.

The law relating to turf cutting is known as *turbary right* and has nothing to do with land ownership. It is a little like grazing rights in the United States. For his $7 fee, which he pays to the Land Commission, Dennis has the right to remove a section of peat seven yards by ten feet. That amount is sufficient to heat his home for one year. Using simple, almost primitive tools, he carves his turf from a bank that was first cut many years ago. In the ensuing time, he's progressed across the bog at a depth of about eight to ten feet.

The cutting, or winning, of the turf, is done traditionally during the months of April and May, or as soon as the crops are planted and the dry weather begins. The basic method has changed little over the centuries. When the bog has drained enough to be worked on safely, the "scraw"—the brushy heather and sphagnum growing on the surface—is peeled back and cut from the top of the bog with a broad spade and a sharp knife. Then, a special lightweight, L-shaped tool called a slane is used to carve away large bricklike blocks of turf. At this point, each piece is so water-saturated that it can weigh up to fifteen pounds.

The turf is some bogs is as much as forty feet deep, and if he wishes, Dennis can cut down to the bottom of the bog. At lower levels, peat becomes progressively darker and denser, and hotter-burning. Dennis always makes sure that his season's supply contains both the quick-burning top turf and the coal-like bottom turf.

At harvesting time, thousands of bricklike chunks of brown-black turf are spread over a couple of acres like so much chocolate on a tabletop. With good weather, the turf will be stiff enough in about two weeks to be piled into little teepeelike stacks known as *footings*. These can better catch the wind to dry the peat at a faster rate. In June or July, the peat usually is put into larger piles for further drying. By the end of summer, the moisture content has dropped to about 30 percent, and the turf has acquired a watertight skin, ready to be brought home as peat by donkey cart, tractor, or truck.

For every ton actually burned to keep the home warm, more than seven must be cut and spread to dry and shrink. Many households use about twelve to fifteen tons a year, requiring one hundred tons to be cut. That adds up to about 180 man-hours overall to cut, dry, transport, and stack the peat at home. Traditionally, the fuel is piled against the gable end of the house to provide insulation, as well as to shelter the peat from rain. [8]

TIDEPOOLS

A tiny tidepool may hold the secret to the causes of poisoning from eating coral reef fish. The little body of water may also increase our knowledge of cancer and heart disease. Where is this tidepool? It's off the coast of Maui, one of the Hawaiian islands.

Researchers investigating a fish poisoning suspected that its cause might be in the food chain. The poison had been passed from herbivorous fish to carnivorous fish. But what had the moss-eating fish eaten? A clue was found in an old book on Hawaiian customs that refered to a poisonous moss in a certain pool found off the cost of Maui. Scientists trying to find the pool ran into resistance from the natives who were obeying an ancient taboo placed on the site by the former kings of Hawaii. The moss was so dangerous, the rulers believed, that it might be put to evil uses. The scientists finally were able to locate the pool, but were warned by the natives that bad luck would follow if they disturbed the moss. In fact, the very day in 1961 that the first collection was made, a fire leveled the Hawaiian laboratory housing the investigation.

The deadly seaweed turned out not to be a moss but a species of soft coral, which was found in one small tidepool only a few feet wide. By means of extremely careful investigations, the coral was found to contain a poison known as palytoxin. The Japanese, a fish-eating nation too, have been on the search for the substance causing vomiting and diarrhea after eating fish from the Ryukyu Islands and off Okinawa. They have decided recently that the poison found in their coral reefs is the same as that contained in the coral from the Maui tidepool, *Palythoa toxica*.

Palytoxin ranks among the most poisonous substances in nature, second only to botulinus and tetanus. What is important in the discovery of the tidepool and its poisonous coral is that the palytoxin it contains will, scientists hope, be a valuable tool in making more clear the complex chemistry of the diseases of the heart and of cancer. In the heart, concentrations of calcium ions are very harmful. Palytoxin perhaps can provide information on how damaged heart membranes take up calcium. In cancer research palytoxin can be used to find out why some cancerous cells are destroyed by the substance.

That small tidepool is giving up its secrets. [9]

ESTUARY

An artist might describe an estuary by drawing the bold pattern of an abalone shell lying on the beach, or possibly his sketch pad would show the circular swirls etched in the sand by beach grass bending in the wind. With a broad brush he could paint the features of an estuary: sand bars, tidal marshes, and mud flats. But how would he show the changing tides, so much a part of estuarine life? It would be as impossible as trying to draw a flower opening or a man growing old. With his camera, a photographer might freeze an estuary's motions: cord grass blowing in the wind, a heron in flight, a bluefish breaking water. But, at best, his photographs can capture but a part of an estuary, a few moments in its history.

A beachcomber might feel differently. To him an estuary is the odd sensation of walking on soft sand, or moving over weed-covered rocks at low tide. His involvement is with the smell of salt air and the odor of the mud flat. A poet or writer seeks mood, inspiration and a sense of unity. An oysterman sees a way of life: an alarm clock at 4:00 A.M. and an aching back from tonging oysters off the muddy bottom.

An estuary means many things to different people. Yet, as varied and unusual as estuaries are, they do have certain features in common. Estuaries form where a river meets the ocean and where the pulsing tide is forever mixing the fresh water from the land with the saltwater of the ocean. Where they meet you may find bays, inlets, lagoons, tidal marshes, sandy beaches, or rocky shores. This lowland area surrounding the estuary is known as the estuarine zone.

The estuary is a place for growing, rich in nutrients and energy. In its shallow, sunlit waters, the incoming tide combines the life-giving nourishment of the ocean with the nutrient-laden waters of the river to produce one of the most productive environments on earth.

"Change" is probably the craziest part of an estuary, for an estuary is an ecosystem in motion, obedient to its masters: the force of the tides, the push of the wind, the gravity of the earth, and the biological demands of all creatures living in it. From moment to moment these forces are at work changing the chemical, physical and biological makeup of an estuary. [10]

SALAMANDERS

Closely related to the vertebrate creatures that first ventured onto land more than 100 million years ago, salamanders belong to the order *Caudata* — the tailed amphibians. Only about 225 species and subspecies are known throughout the world. In comparison, their nearest relatives, the frogs and toads, constitute more than 2,600 species worldwide. Distinguished by their ability to regenerate lost limbs and tails, salamanders are often erroneously called lizards.

Like frogs, salamanders need constant moisture to lubricate their porous skin, as well as a steady supply of insects and worms to eat once they emerge from their daytime burrows. Unlike frogs and other large amphibians, however, salamanders are voiceless and rarely use their feet to seize and hold prey. Instead, they rely heavily on their unusually strong jaw muscles to subdue struggling worms and other creatures.

In the highlands of southern Appalachia, most salamanders begin breeding in the spring, when the males perform a complicated courtship routine. Each kind of salamander has its own unique nuptial pattern. Spotted salamanders, for instance, begin breeding in their Blue Ridge Mountain habitat in March, after the first warm spring rain. Emerging from their winter burrows by the hundreds, they scramble over the ground to their traditional breeding ponds, usually just a short distance away. It is, in effect, a migration, even if only across the length of a valley.

At these ponds, the males and females perform their mating dance, whirling over and under each other in dizzying motion, the water literally boiling with dozens of black and yellow bodies. Within weeks after fertilization takes place, the ponds will be swarming with hundreds of tiny larval salamanders. By midsummer, most of these will lose their gills, develop lungs, and eventually forsake the water for a life on land.

Unlike nearly all its cousins, the marbled salamander waits until fall to breed. At that time, after fertilization, the female deposits her transparent eggs—anywhere from fifty to two hundred of them—in a shallow depression on the forest floor. She then stays with the eggs until the next rain comes, filling in the depression and enabling the eggs to hatch. Fifteen months later, the larvae will be sexually mature.

During winter, only a few aquatic salamanders remain

active in southern Appalachia. With the turning of the leaves, most species begin seeking underground retreats, just below the frostline. There, the tiny amphibians hibernate until spring, when the first warm rains again draw them out to their traditional breeding areas. [11]

SHELLS AND MAN

Most of us have picked up a shell or two on the beach and brought it home. For many of us, that was a long time ago, and the beautiful shell we once treasured has been lost or discarded.

Shells and man have a long, common history. From prehistoric times, their appeal as beautiful, valuable, or mystical objects has intrigued many different civilizations. Shells have been used for money, tools, decorations, and symbols, in rituals, and to make sound. The animals that live inside the shells have been used by man, primarily as food, but also for other things, such as natural dyes.

Archaeologists are able to pinpoint ancient dwelling places by the huge mounds of shells they unearth. Often shells are found far from their normal habitat, giving strength to theories about ancient trading routes. Geologists use fossil shells as indexes to help them date the particular formation from which they were taken. Fossil shells also are used to identify different layers of the earth in geological time.

Cowrie shells, particularly the money cowrie (*Cypraea moneta*), have been used by man for countless centuries. They have been found by archaeologists in many parts of the world—far from the tropical seas where they normally are found. Ancient Egyptians used cowries as a medium of exchange until coins were introduced by the Greeks. The early Chinese made their first metal money in the shape of cowries. Until recently, cowries were used in parts of

Africa as money, and they still are used for personal adornment.

Native Americans used various shells as money. They were known as wampum. Until 1662, when settlers began counterfeiting it, wampum was considered legal tender. Wampum also was used to make belts for ceremonial purposes; it continues to be used as gifts. Returning pilgrims often wore scallops on their cloaks and hats. Others would admire the fan-shaped shells, seeing in them spiritual strength and guidance. [12]

STARFISH

With eyes at the end of each arm, a stomach that can turn inside out, and the ability to regenerate new arms, the starfish—or sea star—surely ranks as one of the ocean's strangest inhabitants.

Starfish, which are found in tidepools of every ocean in the world, are not really fish at all. They belong to a group of sea animals called echinoderms and come in a variety of shapes, colors, and sizes. The largest can measure two or three feet across.

Sea stars get around by way of hundreds of tiny tubelike feet located on the underside of each arm. Suckers attached to the end of the tubes enable the animal to grip onto rocks, as well as to its next meal. Once a sea star homes in on dinner, a scallop perhaps, it locks onto the shell, and its tube feet begin to suck it open. The scallop, with only two muscles to hold its shell tightly shut, quickly tires of the tug-of-war. It's a battle the scallop always loses. After the shell is opened, the sea star pushes its stomach through its mouth, located at the center of its body, surrounds the scallop with its stomach, and digests it outside its body.

A close look at the ends of a sea star's arms will reveal its eyespots, which enable the animal to see only light and dark, not objects. The sea star's arms are expendable, eyespots and all, for new ones can be regenerated easily. Shorebirds or sea otters may take a dive at a tasty sea star and end up with only a tidbit. Scientists have discovered that even a small part of a sea star's arm can regenerate four new ones.

Colorful as they are, live sea stars shouldn't become part of your seashell collection, for they are a vital link in the fragile marine life food chain. [13]

SEA URCHINS

There are eight hundred or so species of sea urchins, and all look more or less like spiny cacti—or living pincushions. Their sharp-tipped, often toxic, spines stick out two or three inches in some species, up to a foot in others. With that unlikely configuration, urchins might seem poorly designed for gathering food but well-equipped for warding off predators. Instead they are efficient diners that all too frequently become dinners themselves.

A sea urchin has not just two jaws, but five. The mouth is on the animal's underside, and its white, pointed teeth are suspended from a five-sided framework called *Aristotle's lantern* because of its resemblance to an ancient Greek light. The whole mechanism is a masterpiece of mechanical engineering, and the jaws open and close in a complex chewing process.

Urchins can use their spines to walk, but their pace is slow—only six or seven feet per hour. The urchin's pace is too slow for catching most sea animals. But on rare occasions a disabled or dying creature—such as the jellyfish—falls victim. Each of the urchin's spines is attached to its body by a ball-and-socket joint, and each can be moved independently. By using its long tube feet, the urchin can pass the jellyfish along these spines until the meal reaches its mouth.

But urchins, which are half an inch across in some species and as much as a foot wide in others, are vulnerable, too. Sea otters frequently dine on them, cracking them open as they do clams. Arctic foxes devour them when the urchins are exposed at low tide. Gulls and other marine birds wolf down large numbers washed ashore by waves.

Numerous fish make meals of urchins, too, but first they have to contrive ways of getting through the spines. The enterprising triggerfish blows the urchin over with a jet of water, exposing the vulnerable underside. The skin of the triggerfish is tough and its eyes are set well back on its head—adaptations that also help to protect it from the urchin's defenses. [14]

BLUE LOBSTERS?

You can tell a dog by his tag, but a Massachusetts scientist knows his lobsters by their colors.

John Hughes' experiment on Martha's Vineyard produces red, blue, striped, and even calico-colored lobsters in an attempt to learn how lobsters can become more plentiful.

Scientists at the Massachusetts Lobster Hatchery and Research Station have been studying normal, plain dark green lobsters for years—in laboratories, and in the wild.

The traditional way of keeping track of wild animals is to tag them, but that doesn't work with lobsters, who shed their shells and grow new ones. Tags would be left with the discarded shells.

To study wild lobsters, something different had to be done. Knowing that fishermen occasionally caught lobsters with unusual shell colors, Hughes asked the fishermen to bring these "freaks of nature" to his laboratory.

He mated the oddly colored lobsters and produced

even odder, more colorful offspring. Hughes soon had a laboratory of brightly hued lobsters. Best of all, when the lobsters shed their shells, they grew new shells of exactly the same color. He had created color-coded lobsters.

In June of 1981, the first of thousands of inch-long baby lobsters were freed along the Massachusetts coast.

The fishermen who catch these colorful creatures will report their finds to Hughes, who will then document how far the lobsters traveled, how much they've grown, and the kind of sea bottom they prefer.

It's too soon to know whether Hughes' experiment will help address problems of depleted lobster populations. Much else needs to be done, such as cleaning up ocean pollution.

But in the meantime, Hughes is trying to help lobsters become plentiful again—and hoping he'll succeed with flying colors. [15]

PICKLED WRINKLES

If you visit in a lobsterman's home Downeast, chances are he will pull a jar of pickled wrinkles out of the refrigerator and offer you one. Wrinkle meat is smooth, pale yellow, and tastes like sweet rubber. It requires a lot of chewing.

The lobsterman may look at you curiously if you ask him what a wrinkle really is. He may be surprised you don't know that people of this northeastern coast have been eating wrinkles for generations. Their ancestors who lived on the shores of England, Ireland, and Scotland ate them too.

A wrinkle is a snail. It is not a periwinkle—that small, thick-shelled mollusk that crawls over rocks and rockweed at the tidemark. *Wrinkle* is a Downeast name for a midwater snail of the whelk family. Other than the periwinkles, the wrinkle is the most common snail in the Gulf of Maine.

Folded away in a wrinkle's mouth is a set of pointed, yellow teeth. When the wrinkle eats, these are extended like the fangs of a tiny wolf. They are perfect instruments for tearing pieces of flesh. The wrinkle moves on a muscle called the foot, a smooth wide band that pushes the animal ahead.

Wrinkles turn somersaults. Scientists found this out by surprise when they were testing the effect starfish tissue has on a wrinkle's behavior. They put a small piece of starfish in the water in front of a wrinkle. The wrinkle's horns retracted instantly. The foot pushed off, hurtling the animal backward. You may have guessed already that starfish, as well as people, eat wrinkles, and obviously wrinkles have known this for a long time. Cod and haddock hunt in schools for these tasty whelks, probing their blunt noses into the kelp forests of the bays.

Probably you have seen wrinkle egg cases tossed up on the shore like old scouring pads. Actually, in the past, fishermen and sailors used them to scrub their hands clean. They called them "wash balls." Inside each capsule, and there are at least fifty capsules to a wash ball, are dozens of eggs. The first wrinkle to hatch begins life with a ravenous appetite. One by one it eats the other eggs. During its two months or more of confinement within the capsule, the little whelk devours every brother and sister. Then it chews its way out of the nursery and crawls free.   [16]

THE LIVING FOSSIL

During their mating revels in the spring, feckless horseshoe crabs may be tossed up on beaches by waves, or left stranded as the tide goes out. In such predicaments, these creatures appear clumsy and poorly coordinated, but a design that has neither been modified nor become extinct in 225 million years must be a good one. Most of these stranded animals will rescue themselves on the next high tide.

In spite of their marine habitat, horseshoe crabs are closer kin to terrestrial spiders and scorpions than to marine crustaceans like crabs and lobsters. Affinities to the air-breathing arachnids include the arrangement of legs, lack of jaws and antennae, physiology of gills and heart, and embryonic development.

Our East Coast horseshoe crab ranges from Nova Scotia to Yucatan and has cousins in Japanese waters and the Indian Ocean.

It takes nine years and dozens of molts for *Limulus* to reach breeding size. When young, the sexes are practically indistinguishable, but with successive molts, the ends of the first pincers in the male take on the appearance of boxing gloves with exaggerated thumbs. The boxing gloves (called pedipalps) of the male fit snugly onto the notched corners of the hind segment of the female, enabling the male to hitch a free ride and be available to shed sperm over the eggs that the female will lay. The hookup can be semipermanent; males are often found clasped to females even during the winter. Adult females grow larger than males, making it easier for them to pull a freeloading male, or sometimes a string of several males, along behind.

Most *Limulus* overwinter offshore in waters at least several meters deep, although some may remain in salt marshes and bays. Various cues, probably vernal warming, increasing day length, and moonlight, trigger the migration into shallower water. In May on the highest high (spring) tides caused by the alignment of sun and full or new moon, females will come so close to shore that their backs are out of water. Each one digs a shallow nest and sheds hundreds to thousands of 2-mm-diameter eggs, beige to gray-green in color, from two pores located on the flap that covers the abdominal gills. While the attending male releases sperm, the female uses legs and gills to create a current that swirls sperm over eggs and causes the whole mass to sink and become covered over with sand. The female then lumbers off, male in tow, leaving a nest that will be marked by a crescent-shaped mound when the tide recedes. [17]

SAND DUNES

What are sand dunes? Sand dunes are the mounds of sand that can be seen lining the landward edge of most sand beaches. Dunes vary in size and shape depending upon wind, wave, and weather, but all dunes are held together by the intricate root system of the dune grass that grows at their surface.

Why is it important to protect sand dunes? First, sand dunes can be thought of as storehouses for sand, ready to replace sand that is eroded from other sections of the beach. if the dunes have been bulldozed or are trapped by seawalls or other construction, eroded sand cannot be replaced. Some beaches gradually are becoming narrow, gravelly strips because dunes are not able to supply fresh sand when it is needed.

Second, sand dunes are an effective barrier that can help save tax dollars by protecting coastal property from destruction. The moveable dunes, unlike artificial barriers, can absorb the shock of storm waves and prevent the ocean from rolling inland to flood or destroy coastal property. After the dune is eaten away by an ocean storm, eventually it is reformed by the natural action of wind and wave. Without the protection of natural sand dunes, millions of dollars have had to be provided to property owners along the coast for emergency disaster relief in the last few years.

What can you do? Simply, make an effort to stay off the dunes. Walking, riding, or sunbathing on a dune can kill the grass that holds the dune together. The grass blades above the ground act as sand traps to catch windblown sand. The intricate root system of the grass locks up sand until it is needed and gives the dune strength and durability. Without the grasses, the dune will disintegrate rapidly, exposing coastal property to powerful ocean storms with no natural defense, and cutting off the sand supply line from dune to beach that keeps beaches wide and sandy. [18]

THE SCIENCE OF OCEANOGRAPHY

From weather to whales, from fluid physics to ecology, the science of oceanography is an interweaving of disciplines that has given us an understanding of our watery planet. Although its pioneers are now retiring, a vast frontier awaits the next generation.

For the student considering a career in science, oceanography offers the opportunity to apply new discoveries to important human and environmental problems. Through the use of high technology to study the ways the oceans interact with the atmosphere, the land, and the seabed, oceanographers of the future will help determine the quality of life and will provide a basis for the laws that will safeguard it. It is a challenge that can yield tremendous satisfaction.

Recently there have been very exciting breakthroughs in each of the major subfields of oceanography. Each is alive with new questions. For example, vertical stirring of ocean waters is thought to be much slower than horizontal stirring, but quantitative predictions still elude us. This uncertainty, of course, casts doubt on the accuracy of certain predictions regarding the disposal of hazardous wastes in the ocean and some important aspects of commercial fisheries management.

Vents spewing very hot water laden with nutrients and minerals have been discovered on the floor of the deep sea. Surrounded by previously unknown organisms, these vents have profoundly affected our ideas about the necessity of sunlight for life-giving processes and even our ideas about evolution. Another puzzle is what appear to be high-energy deep-sea storms, which redistribute enormous amounts of sediment across the abyssal seafloor. These are just a few examples of basic research questions that need answers.

The best undergraduate preparation for an oceanographer is rigorous training in applied mathematics, physics, chemistry, geology, biology, and engineering. Regardless of the subfield a student eventually chooses to specialize in, a broad background in the basic sciences, stressing mathematics, and some experience with computers should enable an oceanographer to follow curiosity as it crosses the traditional barriers between disciplines.

Some oceanographers will go to sea on research vessels in order to obtain the information needed to solve their research problems, while some will get their data in the form of signals transmitted from satellites, stored on magnetic tape, and read by computers. Still others will use the computer for modeling problems. The modeling exercise can be an early step toward determining which data are actually needed to solve a particular problem. What all oceanographers seem to share is an intense commitment to research, an ability to extract the maximum amount of information from any project, and the drive and confidence to complete the tasks they initiate. [19]

SALMON

To those who know it, the Atlantic salmon is much more than a fish. Throughout history it has been a creature that moves humankind, that inspires, mystifies, and intrigues. Twenty thousand years ago, when cave lions and mammoths prowled and plowed through southern Europe, Cro-Magnon man scratched renderings of Atlantic salmon onto reindeer antlers.

The salmon can vault six feet into the air, hit a waterfall, and swim up it. It is acrobatic, powerful, and superb to eat.

But the allure of the Atlantic salmon cannot be explained this easily. Some of the answer may lie in the salmon's ocean wanderings. In a river barren of adults for a season or a year, a salmon will appear, who knows when, from who knows where — a sudden, subtle shimmering seen only, if at all, by those who are looking hard. Where has it been and what has it seen? How has it found its way back?

One navigational tool is its nose. The salmon's sense of smell is so acute that a bear's paw immersed in an upstream pool can bring migration to a complete standstill for five minutes. In one experiment, Pacific coho salmon (close relatives of Atlantics) running up two tributaries of the Issaquah River in the state of Washington were captured and placed in holding tanks. Half the fish had their nasal sacs plugged with cotton wool. Then all the fish were transported below the convergence of the two tributaries and released in the main river. The great majority of normal fish chose the right stream, while the nose-plugged fish made their selections almost totally at random.

But smell alone cannot guide Atlantic salmon from the west coast of Greenland (where most of these North American and European fish winter) to, say, a certain pool on Newfoundland's Upper Humber River. The fish must somehow navigate to within smelling range. Perhaps they plot their course by solar radiation or by earth's magnetic fields. One recent theory holds that gyres — the rotating ocean currents that abound in the northern hemisphere — guide the salmon at least to the proper coast, where they presumably "sniff" until they scent their natal streams. [20]

RETURN TO THE CONNECTICUT

I'm standing by the Vernon, Vermont, fish ladder where, after two hundred years, early this summer Atlantic salmon swam up the Connecticut River past the village of Vernon. It was a memorable event as the fish and game wardens stood in the counting area and watched seven salmon climb forty feet in the recently constructed ladder on their way to the upper reaches of the Connecticut River. They will not be able yet to go all the way to the river's beginning, near the Canadian border, for more dams still bar their way, but work is commencing on ladders on two major obstructions, and in a few years salmon will again be able to swim nearly the length of the river.

The Connecticut, over three hundred miles long, divides New Hampshire from Vermont, splits Massachusetts and Connecticut neatly in two, and empties into the Atlantic at Long Island Sound. It was once so great a salmon river that farmers could feed their families by simply pitching the fish onto the river bank with a hay fork. Hired men, it is said, made it a part of their work agreement that they should not be fed salmon more than once a week. Then, in the late 1700s, dams were built across the Connecticut, and salmon and shad (among other fish) were blocked from their ancient spawning grounds.

Salmon are a migrating fish; they leave their birthplace in the rivers and once in the ocean travel 2,500 miles to the Greenland coast where they mature. Then the urge to mate forces them back to the waters they knew as smolts. Now that ladders have been built around many of the dams on the river, the adult breeding salmon will swim up stream to suitable spawning places in the tributaries. The next generation, a few years hence, will return to the spot where the water, the river bottom, the light seems just right, just the place that they knew as baby salmon, and they will in their turn spawn.

It is hoped that the Connecticut River will again be known as one of the great Atlantic salmon rivers. [21]

THE SEAHORSE

Ludicrous, comical, and inappropriate in appearance for a fish, seahorses have somehow survived and thrived through millennia.

Seahorses are, in fact, beautifully adapted for the lives they lead. The pipelike snout, for example, is a rigid vacuum hose — if an amphipod or other tiny crustacean appears, the seahorse sucks it in more quickly than the eye can follow. In stalking unobtrusively about marine grasses and seaweeds, they seem to drift aimlessly, yet they are propelled by extremely rapid beats of their almost transparent dorsal and pectoral fins. Swimming upright, they resemble bits of drifting seaweed, a similarity enhanced by the long fleshy projections found in some species.

There are about thirty species of seahorses, ranging in size from one and one-half-inch pygmy seahorses (*Hippocampus zosterae*) to giants nearly a foot long. Some seadragons, relatives of the seahorses, grow to eighteen inches. Although very similar to seahorses, the seadragons have their heads set at more oblique angles to their bodies.

Both seahorses and seadragons have prehensile tails that curl ventrally. With these, they cling to plants, rocks, and sometimes one another. Instead of scales, seahorses and their relatives have bony plates girdling their bodies.

Probably the best known and most fascinating adaptation in the seahorses and their relatives is the pouch, or brooding site, possessed by the males. The females deposit their eggs in the pouch of the male, the eggs being fertilized during deposition. One or more females may make several deposits before the brood pouch is full, whereupon the female is relieved of further responsibility. The eggs are brooded by the father until the young hatch and are ready to swim away. The father then expels them, several at a time, by contractions of the pouch and flexation of his trunk. Sometimes an infant dies in the pouch, and the gases of decomposition make bubbles that float the father to the surface. Here he will die unless the bubbles are squeezed out. Up to several hundred eggs may be brooded, depending upon the species, and the father takes no further care of the infants after they are born. [22]

SHRIMP FARMING

There are few finer seafood experts, especially when it comes to shrimp, than the Japanese.

The high demand for shrimp in this densely populated country long ago caused the depletion of shrimp in the ocean around Japan. Fishermen, using ever more sophisticated techniques, were forced to look farther and farther from Japan for shrimp.

But the costs of fuel, boats, and labor continued to rise and so did the price of shrimp. Some Japanese businessmen soon realized that it would be easier and cheaper to raise shrimp in ponds all the way to adults.

One of the main problems of high-density shrimp culture was disease. Because of overcrowding, the pond bottoms frequently became saturated with waste products, which led to sharp decreases in oxygen in or near the sandy bottom. Shrimp instinctively spend much of their time burrowed in the sand. Many of them, under such conditions, became weak. Infections, especially in the gills, were common in shrimp ponds causing slow growth, poor quality, or death among the shrimp.

Starting in the early 1970s, a team of researchers developed an entirely different approach to the culture of young shrimp. They built a large, concrete tank almost one-fourth acre in size. It was circular with a drain in the middle. Water entered the tank from a series of nozzles along a pipe running from the edge to the center of the tank. This caused the water to swirl, improving the circulation of oxygen and new water to the bottom of the tank. Anyone who has ever swirled water in a bucket will also realize that this new design automatically swept pieces of dirt and shrimp waste products toward the center drain.

The new design decreased disease problems sharply. Rumors soon spread in the Japanese shrimp industry of the new tanks capable of producing over 17,600 pounds of shrimp per acre per year, four times what was previously possible.

Many shippers now pack live shrimp in moist, cool sawdust. Over short trips (less than twenty-four hours), the sawdust prevents the shrimp from drying out, and it reduces injuries.

In Japan, intensive methods of shrimp culture are, therefore, beginning to provide quality shrimp, on a more consistent basis, at a cheaper price! For Japan, the dream of shrimp farming has become a reality. [23]

Places

CANYON DE CHELLY

Our yellow maple leaves are magnificent every fall, but I must admit that some of the brightest colors I have ever seen were far away from New England, in northern New Mexico, Canyon de Chelly, to be exact. Canyon de Chelly is a half-mile-wide valley bottom whose absolutely sheer sandstone walls soar straight up for a thousand feet. It is probably the closest place to paradise on this continent. For in this desert country, the canyon has a winding creek, the soil is fertile, it is tree-shaded, and above all it is protected from storms.

Ancient Indians who lived here before Columbus arrived on these shores had a little civilization here below the surface of the desert. They built sophisticated apartment houses in the shallow caves under the cliffs, apartment houses that had a system of air-conditioning and central heating. They vanished centuries ago, and now the Navahos, no relation, farm the beautiful valley.

We walked one fall afternoon down into the sandstone canyon, which showed all shades of pink, lavender, and cream. At the bottom the shallow stream wandered from side to side, gray willows and brilliant yellow cottonwoods marking its course. Before us were the ruins of the famous so-called White House about seventy-five feet above the floor of the canyon, tucked into an oval cave, its protecting roof the thousand-foot cliff that rose to the level of the surrounding desert above.

In the late afternoon quiet, in the beautiful sunny valley we could sense the pride of the people who had lived here. They had left their pictographs on the rock wall of the canyon, a deer painted in white against the pink stone, a man lying on his back playing a flute. Why did they leave this paradise? Because of a drought? Where did they go? Their roofless house gives no answer. But we had the feeling that somehow their happy spirits still inhabit the lovely Canyon de Chelly. [1]

EARLY PRAIRIE HOMES

The first prairie settlers, with free choice of land, had no lack of sound building materials. They certainly had no lack of experience; they simply built as they always had, hewing, shaping, and fitting with practiced ease. If there were eight or ten good men on the job, a cabin could be raised and roofed in only a few days. The fewer the builders, the longer it would take and the more flimsy the finished result was likely to be. But with plenty of help to "carry up the corners" of heavy log walls, the cabin would be a solid and enduring home. Some early log cabins along the prairie frontier were continuously occupied for seventy-five years.

If there was time, the hardwood logs of the cabin walls might be squared with broadaxes. Pine was seldom available, of course, and the walls were often made of white oak—although some early cabins were made almost entirely of black walnut. There were rarely any nails or other hardware available; joints were cleverly mortised and tenoned, or secured with pegs of black locust driven into auger holes. A roof was made by laying very straight small logs from gable to gable, and overlaid by clapboard slabs nearly five feet long. Weight poles were then laid over the roof, again secured with wooden pins.

Fireplaces and chimneys presented a special problem in that land with more birds' nests than stones. Some flues were made of sod plastered inside with clay, but the most common type was the "cat and clay" chimney made of small split stakes about three feet long, thickly plastered with clay. In cold weather, such chimneys frequently burned. The fireplace itself had to be built of stone — or of stone and clay, or wood faced with stone. In any case, it was usually planked on the outside of the cabin with short butts of logs to contain the rock and mortar.

The finished cabin might have a puncheon floor made of split logs, with storage chests and boxes kept under high bunks. In real frontier society, the squared inner walls of the cabin would be whitewashed, and the actual cooking and eating might be done in a smaller cabin a short distance away from the main house. The typical cabin, however, combined all the family activities in one room and perhaps a loft. The walls were lined with pegs for spare clothing and utensils; from the rafters hung sides of bacon, smoked hams, venison saddles, and rings of dried pumpkin and squash. And always, just over the door, were hung rifle, powderhorn, and bullet pouch.

Building materials for houses in the 1840s grew progressively poorer farther out on the open prairies. Settlers there were denied oak, hickory, and walnut and were forced to rely on cottonwood, ash, and even willow. They would build cabins of cottonwood logs that were roofed with willow and sod, and in extreme cases they might even build family cabins entirely of willow poles. But in the outer reaches of the tall prairie even those poor building materials weren't to be had, and the settlers were forced to make homes from the prairie sod itself.

Sloughgrass sod was the best, with its dense matrix of strong interlocked roots. Lowlands of sloughgrass and big bluestem were scalped of their sod, which was usually cut into slabs a foot wide and two feet long. The course of a wall was laid two sods wide, and the next higher course would be laid at right angles to the first. This sealed all joints, and the wall of a sod house was a weatherproof two feet thick—cool in summer and warm in winter. The roofs, made of pole framing, were covered with thin turfs and thatched with the tough sloughgrass. An elderly Nebraska lady once told me that the roof of her sod home "leaked two days before a rain and for three days after."

A fortunate prairie woman might have a soddy with whitewashed inner walls and perhaps, later on, even a couple of wooden-sashed glass windows. But rarely, if ever, would there be anything but a dirt floor.

Many soddies were never much more than miserable, depressing hovels, but a few were grand, two-storied, shingle-roofed affairs that might even have trimmings of Victorian gingerbread—about as wild a concession to both function and fashion as the prairies ever saw. As a rule, however, early settlers rarely took any pride in their soddies. Old wet-plate photographs of farm families outside their sod houses often included a few prized possessions that had been hauled outside for the occasion: a new organ, a canopied perambulator, and perhaps a table and chairs that had been "brought out from home." It was their way of letting the world know that they were getting ahead, even though they might live in a cabin made of earth.   [2]

YELLOWSTONE BACKCOUNTRY

The leader of our hiking party has a bear bell on his pack, worn in the hopes of scaring off any wandering grizzly. We have just taken off our packs, for we are now in our assigned camping spot for tonight. No one is allowed into the backcountry of Yellowstone Park without permission, and hiking parties must stay only at the campsites that have been allotted them.

The first thing we will do is pick a spot for our tent, one that seems the smoothest and that will have the morning sun on it when we arise, for getting dressed in twenty-eight-degree morning chill is one of the less pleasant aspects of hiking. Then pads are laid inside with the sleeping bags on top all ready to be slid into. The next thing is food, and on this trip cooking is easy. Our outfitter and guide has provisioned us with freeze-dried food, all labeled, first lunch, first dinner, first breakfast. Even though it's freeze-dried, it still mounts up to sixteen extra pounds that each of us has to carry for the nine days that we'll be out on the trail, in addition to our clothes, sneakers for fording rivers, sleeping bags, and tents. It's more than most of us have ever carried, and we're only happy that we are hiking on the relatively easy grades of horse trails and not on the steep, rooty trails of our native New England.

As I said, the cooking is easy; the little backpack stoves are lit, pots of water set to boiling, and we simply slit open the pouches, pour in a cup or so of hot water, stir it up and let the dinner hydrate for ten minutes or so, then eat out of the pouches. In the meantime, we drink a freeze-dried soup, mixed with hot water, from our cups. We carry only a cup and a spoon, so cleaning up chores are minimal. Of course, we never wash them out in the cold streams, but instead simply scoop up some water, rinse out the cup and spoon, and throw the dirty water away from the stream.

The food is filling; you need lots of starches and sugars on a hike like this. Tonight we're going to have mushroom soup, chicken Tetrazzini, and chocolate pudding. The empty aluminum pouches we simply roll up and pack out. The campsites we've seen have been well cared for by other hikers, who are thoughtful enough to leave some campfire wood for us. We'll sit around for an hour and compare notes with our guide, whose life in Montana seems so different from ours.

But the main activity in the evening after dinner is the process known as hanging the packs. Grizzly bears are unpredictable creatures. They may run from a hiker or, if startled and guarding a cub, they've been known to attack. Park policy regarding them has changed. It used to be that people gathered at park dumps to watch the bears feed. They came to depend on human handouts, a very dangerous situation. That has stopped; the bears are on their own again. But their keen sense of smell will lead them to food. Perhaps freeze-dried food wouldn't have much of an odor, but we will take no chances. Even our toothpaste will be hung. A suitable tree some distance from the campsite is chosen. A rock tied to a rope is thrown over a limb about ten feet high. The food-stuffed pack is then hoisted into the air for the night. The woods around our site will look odd with fifteen gaily-colored packs strung up in the air like ungainly Christmas decorations.

Now the darkness is beginning to settle around us, sweaters are pulled on, and the campfire is lit. Later from a far hill perhaps the lonely wail of a coyote will put us to sleep when we have crept into our little tent. [3]

MOUNT ASSINIBOINE

By April most people are rejoicing in the signs of spring. In the high reaches of the Canadian Rockies, however, winter still has that high world in its icy grip. Snow mounts higher, glaciers inch down mountainsides, avalanches follow their deadly course.

In April a few years ago Jane and I decided we were not ready for spring and with some other adventuresome souls met our guides in Banff, Canada. We were going to cross-country ski in Mt. Assiniboine country. Mt. Assiniboine is one of the world's most beautiful mountains, but its pyramidlike peak is seldom seen except by those who venture into its fastness.

We were each given an electronic beeper to wear to aid in finding us if caught in an avalanche, and off we set on our skiis, packs on back. It was a ten-mile trip, above the tree line, and a trifle frightening because the year before a party on the same route had been caught in an avalanche. But the sensation of skiing on top of the world soon caused us to forget our fears. On top of Sentinel Pass we were ferried by helicopter over an awesome wilderness to a cabin at the foot of Mt. Assiniboine.

The cabin was almost totally buried in snow, but we lived there comfortably for a week. Warned never to leave the cabin without our beepers and glacier goggles, we set off every day, lunches in our packs, to ski with our guides in the heavenly fresh powder snow that came up to our knees. There was always something to see, a lynx track leading to a snow ptarmigan's hole in the snow, or a golden eagle soaring overhead. Our guides, expert mountaineers, led us on the routes around the mountains that were most likely to be safe, but there was always the fear that the slope one was crossing might give way, or an overhanging cornice above would come crashing down.

Usually we forgot about avalanches in the delight of the beautiful pristine, blue and white world, except for the show that Mt. Assiniboine put on every afternoon. We would gather outside the cabin to look down the frozen lake at the mountain. Soon there'd be a cracking sound, and in a second, with a tremendous roar, the sun-warmed glacier would send down blocks of blue ice big as houses. We were glad to be at the other end of the lake.

We hated to leave this wonderful place, but back again in Banff, we heard a robin sing and knew we were ready for spring. [4]

THE COLUMBIAN GROUND SQUIRREL

We were eating our trail lunch in one of the Canadian Rockies' high mountain meadows. The little grassy valley bottom with its winding clear stream was a perfect place from which to see one of nature's most impressive shows: avalanches. Slipping off the thousand-foot cliff of the mountain opposite us, the snow fell like a waterfall. Suddenly, above the intermittent roar and rumble of the snow and rocks, we heard a sharp whistle nearby and turned to look for an approaching hiker. No one was near, but in a bush close by we saw a little furry animal sit up and peer at us. We sat quietly as it approached, then retreated, in short rushes until at last it came close enough so that we could see it was about thirteen inches from head to bushy tail, with dark, reddish feet and underparts, and with a beautifully mottled gray fur in its back.

We later learned that it was a Columbian ground squirrel. It lives on grasses, and in the fall stores a few seeds, in its range of British Columbia, Idaho, and Montana. From autumn to March it goes into deep hibernation, when the two to eleven young are born in a special underground nursery. Deciding that we were neither of his natural enemies, hawks or coyotes, the little fellow grew more bold and accepted a precious bit of salami from our fingers. One taste convinced him it wasn't what he'd hoped for, for he refused another piece. But when an edge of a heavily spread peanut butter sandwich was offered, it was snatched away so greedily that the whole sandwich was torn out of the offerer's fingers.

That evening back at our tents a biologist told us that the ground squirrels are now being studied for their social behavior and that they are almost as varied in their personalities as humans. Placed before a mirror, some Columbian ground squirrels are aghast at their appearance and flee from the sight of themselves; others attack what they consider an enemy, while still others preen and primp at their image. Our attractive little animal living in his high green valley will never see his image in a mirror, but for us humans he had a definite personality that made the lovely valley even more beguiling. [5]

CARIBOU

"You should have seen them, the caribou and the wolf pack," the young park warden said. "There they were, a whole herd of caribou, grazing as they went, and following them a pack of wolves." The warden paused as she remembered the scene she had witnessed only a few months before in the late spring in the same valley where we were. We had hiked fourteen miles into the back country of the Canadian Rockies to the foot of Mount Robson, Canada's highest peak, and had stopped at the warden's cabin to speak to the young woman, who was also a naturalist. The scenery was magnificent. From the desolate valley with its ice-blue lake rose Mount Robson, soaring another six thousand feet into the clouds, its sides shining with glaciers and snowfields.

The park warden had seen the annual summer migration of the caribou, who were returning to their summer grazing grounds in the high mountain meadows where the grass is surprisingly rich and green. The following wolves were not harassing the herd, just following, hoping for a stray elder who had been weakened by the winter's travails, or a sickly youngster.

These were mountain caribou, bigger and darker than their cousins who live in the Arctic. Both are related to the European reindeer. Well adapted to their brutal climate, their fur is hollow and air-filled, with a thick woolly undercoat, and even their nose and pads are covered with hair. Their hooves are rounded and spread when walking over snow and can be used as snow shovels to dig out the mosses and lichens they live on in winter. Both sexes grow enormous antlers that sweep backward. The world's record was a pair found in northern Quebec that ran to a length of seventy-four inches. The young are born in late spring after a seven and a half-month gestation period and can run after being in the world only two hours. A day-old calf can run faster than a man and in three days can keep up with a galloping herd. The wolves she saw following so hopefully would by no means be assured of an easy meal.

We think she was glad that she hadn't seen a baby killed but, as she said, winters there in the mountains are very long and extremely cruel. The outlook for an ailing caribou or a sickly calf is starvation, and a quick death by the wolves must be preferable. The back country of the Canadian Rockies is magnificent, but existence there is exceedingly difficult. [6]

ADIRONDACK GUIDE BOAT

We remember well our expedition in an Adirondack guide boat on the choppy waters of Lower Ausable Lake. The lake is one of hundreds in the largest public park in the United States. Adirondack Park in upper New York State is larger than Yosemite, Grand Canyon, Olympic, Yellowstone, and Glacier National Parks combined. Lower Ausable Lake is an awesome body of water, two miles long, from whose narrow length cliffs rise a thousand feet. After pushing off from shore, there is no place to land, no friendly beach, no trails, only those fearful, soaring rocks. We had to keep going until the carry at the upper end that leads to another lake. Although the waves were white crested, we had complete confidence in our frail little craft.

She was a beautiful, elderly lady, almost eighty years old, sixteen feet long and weighing a mere sixty pounds. Her thin skin was of cedar fastened with roughly five thousand copper tacks and three thousand brass screws. Very shallow in the center, an Adirondack guide boat has sheer bows and stern with a conspicuous rise to throw off the splashing whitecaps. These boats are rowed with long oars, so long that the oarsman has to cross one hand beneath the other. The long oars propel the light craft with great speed, almost like a racing shell.

These remarkable boats were developed in the 1830s by the woodsmen of the region who needed something sturdier and faster than the birchbark canoes of the Indians. By 1900 they had reached the height of perfection, the loveliest of all watercraft. Able to carry three people with laden packbaskets and perhaps a slain deer, yet light enough to be easily portaged over the carries that link the miles of lakes, they have given way to mass-produced boats that are squeezed out like toothpaste from a tube.

Still, there are lakes and streams where motors are banned, and some proud owners again launch their beautiful boats, each a little different. Our eighty-year-old lady proved on our trip that she was a true aristocrat of watercrafts, an Adirondack guide boat. [7]

THE CARDIFF GIANT

While traveling in upper New York state, Jane and I found that there always seems to be someone around who hopes to make an easy dollar by fooling his or her fellow citizens. Everyone knows the story of the Brooklyn Bridge, the city slicker, and the country bumpkin, but some may not remember the name of George Hull, the Binghamton, New York, cigarmaker who perpetrated one of the biggest hoaxes on the American public (so far, that is).

George Hull's theory was that people actually wanted to be fooled if a hoax bolstered local pride or promoted a popular idea or cause. In 1869 new scientific ideas in geology led a craze in collecting fossils. What better, thought George Hull, than to dig up a really spectacular fossil like a man? P.T. Barnum had made a fortune exhibiting a huge elephant, so maybe a fossil man would be as popular. Accordingly, he found a partner who helped to locate a gypsum mine in Iowa. A twelve-foot slab was sent to Chicago where two sculptors, sworn to secrecy and provided with all the beer they could drink, turned out a statue of a man. The trouble was, it looked too new, so Hull poked it all over with knitting needles and doused it with ink and sulfuric acid to give it a patina of age.

One other man was let into the secret, Hull's cousin, a man named Newall who lived near Cardiff, New York. By roundabout back roads, the stone man was carried in a big box in a wagon to Newall's farm, then buried at night behind the barn. Hull was cautious and waited for a year before he told Newall to hire men to dig a well just behind the barn. To their great surprise the diggers came upon the body of a twelve-foot petrified man!

On the first morning, neighbors crowded around the hole, and by the next day Newall put a tent around the excavation and charged $.50 to see the wonder. In a short time, as many as twenty-six hundred people on one Sunday paid their money to enter a small darkened tent. There the spectators leaned forward to see the huge naked petrified man in his grave. Hull thought it best to take his profits quickly before the fraud was discovered, so he sold the giant to a group of businessmen from Syracuse for $30,000. Syracuse was proud to have such a remarkable wonder and heartily refuted any suggestion that the giant was a hoax. Even after it was proven that it was, people continued to believe in the Cardiff giant. Today he rests in Cooperstown, New York. [8]

WINE FROM THE FINGER LAKES

We stood in the tasting room of one of the many wineries in New York's Finger Lakes Region, watching the young woman who had taken us through the winery, pour from a bottle of white wine. "It's a Seyval Blanc '79; tell me what you think of it." We held our glasses to the light, sniffed at the aroma, and took a tentative sip. "Dry and crisp," said a knowledgeable man.

Through the windows we saw gatherers moving along the rows of grapes that stretched down the steep hillside to the narrow lake. "It might be the Rhine Valley," we remarked, "just needs a ruined castle or two." Our guide agreed and told us that indeed the Finger Lakes area does have much in common with Germany's wine growing area. It has the same shaley soil that looks as if nothing would grow, but whose minerals supply just what the finicky grapes need to produce excellent wine. The lakes act as the Rhine does, to ameliorate the climate.

As we watched, tractors towing one-ton bins piled high with grapes unloaded their juicy burden into pressing vats. The resulting mush was drawn off into oaken fermenting casks to wait for seven to fourteen days until the winemaster decides the process is at exactly the right state. The must, as it is now called, is then racked off into other casks where the sediment will have a chance to settle. We followed our guide down into the cellars, inhaling the delightful winey smells, and watched a skinny man squeeze into an enormous cask through a tiny hole at the bottom, to clean out the cask. Because of the heady fumes he would remain there only twenty minutes.

At the last stage of winemaking, the winemaster must employ all his skills to determine just when the wine is right for bottling; the winery's reputation rests on his decision. This is the culminating point of the year, a year that began in the winter when the workers went out into the vineyards to prune the thousands of acres of vines. In spring trellises must be repaired and the vines tied up, and in summer the cultivation must be carried on. Finally, in September, the early grapes, the Seyvals, a new hybrid, are picked, and last the native catawbas in late October. In New York's Finger Lakes Region, the ancient art of winemaking is indeed flourishing. [9]

ST. BRENDAN

We stood one rainy afternoon on the coast of the Dingle Peninsula in County Kerry, Ireland. From a cliff we watched our fisherman friend Eddie, at whose house we were staying, launch his curragh in the tiny rock-strewn harbor. A curragh is a frail craft, canvas stretched over an oak frame, rowed with long polelike oars. It's on this coast that the legend of seagoing St. Brendan persists, he who sailed in a curragh very like Eddie's and, the story goes, first found North America around A.D. 500. St. Brendan's tale is full of magical events but his course, an island-hopping one, can be traced.

In 1977 Englishman Tim Severins took some ox hides, sewed them together with linen thread, and put them over an oaken frame. Then, with four others, he sailed from the Dingle, up the Hebrides, to the Faroe Islands, to Iceland and Greenland, landing in Newfoundland. He kept the curragh watertight with lanolin and once fixed a hole by sewing on a leather patch. "We would have drowned if the boat had been of fiberglass," he said.

Other modern materials also proved unsatisfactory. Down sleeping bags were wet, artificial cloth cold, dehydrated food disintegrated. In Iceland the crew stocked up on woolen clothes, smoked meat, and cheese and nuts, the same sort of supplies St. Brendan might have used. Severins made it to the shores of North America in his unlikely craft, a survival of the Stone Age, and perhaps St. Brendan did too, with his fourteen monks, one thousand years before Columbus. Anyway, it's fun to think so.   [10]

FORGOTTEN VIKING BOAT

Not many people have had a chance to lay a hand on a Viking boat—the existing craft are too well guarded. Well, I have touched the gunwale of a Viking longboat, actually not with my fingers but with an oar, for the boat lay twelve feet below me in the water.

I was fishing with a companion on the waters of Lough Corrib near Galway, Ireland. It was lake salmon we were after, and when we'd caught a few it was time for lunch. As we reeled in our lines, my host turned his stout boat toward one of the many islands that dot the water of the twenty-seven-mile-long lake. On shore he got out his little parafin (we call it kerosene) stove, and we boiled eggs and brewed tea. There was also some good Irish cheese and soda bread.

After we'd eaten, he said, "I'll show you something because you're an American, and if you tell it won't make any difference. I don't want people around here knowing of them." I was consumed with curiosity, but he said nothing more while he carefully piloted his boat looking for bearings. Finally he stopped and told me to look down. Through the clear water I could see the shadowy outline of a half-buried boat. It was a Viking longboat about twenty-six feet long.

The Vikings had long ago made a lightning raid on the unprotected Irish coast, then marched inland to raid some more. The enraged Irish defeated them in a forgotten battle, and the Viking chief ordered that the boats be sunk in the lake, intending to return and raise them once more to beat the Irish. But the Vikings never returned, so to this day the boats lie, half buried in the sand, their rudders nearby awaiting their fierce masters. One thousand years later I leaned over the side of my host's modern boat and touched with an oar the gunwale of a long-forgotten craft. [11]

BEATRIX POTTER

Most of us had the *Tale of Peter Rabbit* read to us when we were growing up. In fact, Peter must be one of the best-known characters in English literature. His creator, Beatrix Potter, was the daughter of a wealthy, domineering Victorian papa who kept his talented daughter firmly under his thumb. Her only outlet for her artistic drive was in studying with great care the little animals she kept as pets, and in using them as models for the illustrated stories she wrote for young acquaintances. By a fluke, her little tale of Peter Rabbit was sent to a publisher and was an immediate success.

Greatly daring, she braved her father's displeasure and with the proceeds from the book bought a small farmhouse in the Lake District in the north of England where she had spent carefree holidays as a girl. Here, she escaped when she could from her parents' stuffy life in London and wrote the wonderful stories children all over the world have loved.

Finding ourselves near the tiny hamlet of Sawrey, where her house is, we felt we must pay it a visit, although Beatrix Potter has been dead for some forty years. She never allowed any modern heating or electricity in her house, but as we stepped into the stone-flagged kitchen heated by a fire in the old-fashioned range, the dark cupboard bright with china, we sensed her warm and loving spirit, which filled the room so that it needed no modern convenience. It all seemed familiar, and of course we'd seen it hundreds of times before—the clock, the newell post, the armchair—when we'd looked as children at the pictures in the *Tailor of Gloucester* or in *Benjamin Bunny*.

Beatrix Potter finally married and lived happily as a sheep farmer, when she realized that her beautiful countryside was in great danger of being spoiled by tourism. She became convinced that the only safety lay in land donations to the National Trust, and during her lifetime she turned over four thousand acres to the Trust, ensuring that the exploiter and the jerry-builder would not threaten England's lovely lake country. We felt that we owed Beatrix Potter a double thanks, for Peter Rabbit and for her work in nature conservancy. [12]

LAPLANDERS

I thought recently of the strange Laplanders we saw in Arctic Scandinavia. We were fascinated by these unusually dressed people, who seemed to be of a different race than the other Scandinavians. Their language was different, not understood by the Finns and Norwegians. They were short and had broad faces with pointed noses and chins. Where had they come from?

Anthropologists now think that the Lapps may be survivors of the last Ice Age people, hunters of the glacial age animals such as the woolly mammoth. When the great glaciers started to melt, the animals retreated with them followed by these hunters, who settled in the barren arctic lands of Norway, Sweden, Finland, and Russia. The mammoths disappeared, so the hunters turned to the reindeer as their means of livelihood. Reindeer are the Lapp's source of milk and of meat. They use them to transport their portable houses in summer and to pull little sleighs in winter. Not all Lapps are now reindeer-herders; some have settled into fishing villages, and some work in the forests.

But all dress in the traditional costume. We had never seen anything like it before, but somehow it looked familiar. Then we realized it was because the Lapps still are dressed as people in the Middle Ages. The men's dress is the only medieval dress outside a museum or a picture book. It is a dark blue tunic, brightly decorated, belted low on the hips, with blue leggings, fur boots with turned-up toes and a fantastically shaped cap, different in the various parts of Lapland. In the mountains, tundra, and bogs, these people wander across the boundaries of four countries, for borders mean nothing to them as they follow the reindeer.

The Romans knew of them as strange people who lived where the sun never rose, pursuing animals by skimming across the eternal snow on pointed boards. Two thousand years later we found the Lapps living in much the same way as their ancestors, those ancient folk who hunted the woolly mammoth at the edge of melting glaciers. [13]

THE LIPIZZAN HORSES

One of the most remarkable sights in this modern age of ours is that of the white Lipizzan horses performing battlefield tactics of the eighteenth century. It seems so incongruous; here we are in the nuclear age when a touch of the finger may send us all into dust, when the gallant arts of war are as dead as the proverbial dodo, yet thousands of spectators sit spellbound for hours watching the Lipizzan horses perform cavalry exercises that were designed four hundred years ago.

Why do we want to see this? Perhaps it's because today we have such few chances to witness communication between animal and man. We saw the magnificent stallions in the dead of winter in Vienna, when all foreigners but us had left for home. On a frosty January morning, Jane and I walked into the Spanish Riding School, home of the white horses for over two hundred fifty years, a palace ballroom lit with glittering chandeliers, but with sawdust on the floor. We two, the only audience, sat for hours blowing on our numb fingers, while men and their steeds went over and over the incredibly complicated routine. We never thought of leaving. The patterns they wove, the marvelous discipline, was fascinating even to our untutored eyes.

The riders were dressed in uniforms of the Napoleonic age: brown tailcoats, black cocked hats, white leather breeches, and knee-high jack boots, and their seat, or posture, on the horse's back is far different from that of modern-day riders. So close is the communication between that of rider and steed that to us the aids were unnoticeable. One

Places   219

of the most spectacular figures was the capriole, where the horse leaps three feet straight up, at the same time folding his forelegs against his chest while kicking out the hind legs so that his shoes can be seen.

Later, as we traveled down through Austria on our way south to Italy, we saw the herd of Lipizzaner mares grazing in the mountain pastures with their foals, who are black at birth, becoming white only as they mature. Beautiful horses with skilled riders are fascinating in a world where beauty and skill are becoming ever more rare.   [14]

THE TOWER OF PISA

They say the Leaning Tower of Pisa is leaning even more these days; it tipped five one-hundredths of a inch last year, and Pisa city officials say its doomed to collapse unless a way is found to halt the tilt. It may be one hundred years before it collapses, or it might be much sooner.

I'm glad they didn't tell us that a few years ago when we were there, or we wouldn't have walked under it, much less actually climbed the tower. But we didn't know it was leaning more every year, so we paid our liras and started up the marble steps worn to icy-like smoothness by eight hundred years of climbing feet. Jane hasn't much of a head for heights at the best of times, and it wasn't long before she wished she'd stayed firmly on the grass below with the more prudent members of the party, particularly when we had to negotiate the downward side of the tower. We had the sensation that it would be our weight that would finally send the whole mass to the ground.

But we kept on, Jane very slowly, all the while clinging to the innermost side, her arms hugging the wall. Then she encountered another outstretched hand, a male hand and squeezed around to see who it was who also was so frightened. "I'm scared to death," said Jane. The man smiled a pale smile. "So am I, and I'm a British Airways pilot. Flying a 747 seems a lot safer than climbing this crazy tower. I'm sure it's going to fall in the next few minutes, or that I'm going to slide off on the downward side." Well, we all made it to the 180-foot top, although I think Jane had her eyes closed the whole time she was there.

No one knows why the tower, begun in 1173, leans. When it was thirty-five-feet tall, it was already ten feet off the perpendicular. Engineers know that the tower tilts faster when the water pressure below is weak, and remains steady when it's high. Last year was dry, and perhaps that's why it tilted more.

We're glad we climbed to the top, and we can only hope that the ingenious Italians will find a way to keep the Leaning Tower of Pisa from leaning too far.   [15]

SARDINIA

We found when visiting Sardinia that it is one of the world's large islands and right in the much visited Mediterrean, but very few Americans ever go there. It's an island with its own culture left by the Phoenicians, Vandals, Greeks, Romans, and in the Middle Ages the Spanish, who once owned it. It's been little touched by the twentieth century as we found the night we boarded the boat from the Italian port. An amazing buzzing noise came from the stern where a group of returning Sardinian shepherds were singing native songs; to us they sounded like a hive of angry bees. We later learned that this form of music may be as old as Stone Age man.

The coastline had the most beautiful beaches anywhere in the world. The rocks and cliffs looked as if a giant had been blowing bubble gum, the result of ancient volcanic gasses popping and exploding in the lava. Behind in the rich interior, six thousand-foot mountains rear snowcapped peaks. Eagles soar there, and wild boar and moufflon, a European mountain sheep, live on their steep sides. In the little mountain villages the people still wear traditional dress, beautifully embroidered shawls for the women, and for the men, white pantaloons, a black cap with a long flap, and long black gaiters.

The countryside is rich in ancient remains, elaborate graves that date from the time of Stonehenge, and most particularly the strange nuraghis. No one knows who built them or what they were used for but they date back to the Bronze Age. They're truncated hollow towers built like a cone and with an elaborate system of stairways inside the double walls.

Sardinia is sparsely populated compared with mainland Italy, and its people appear prosperous with their olive groves, cork trees, and fruit orchards. In some of the most remote towns the banditti still hold sway, carrying on happily their ancient feuds by rustling sheep and kidnapping one another. The Aga Khan has brought a new note of commercialism to Sardinia by building some resorts for the wealthy along the sea coast. The owners were surprised that anyone would want this coastal land, for it was the interior that was prized. Now they are abandoning their simple ways while their inland brothers regard them with jealousy, and feuding is common. [16]

CORDOBA

Who was it who said that Africa begins in Spain? The minute Jane and I left the French train at the border to board the Spanish train we felt we were in another continent. (France and Spain have different railway gauges and can't use one another's tracks.) Before we were given permission to board, our luggage was searched by the *guarda civile*, in uniforms topped off by a distinctive flattopped leather hat. Carefully they drew on white gloves before going through our bags, then waved us on with a weary bored gesture.

After an interminable wait, the train started forward with a jerk and slowly gathered speed but not too much, for the birds seemed to be flying faster than we. The landscape looked different, dry and bare; we saw flocks of sheep tended by a motionless shepherd with a crook. We followed the dining car attendant to the restaurant and had something called *calimar zarzuela,* which turned out to be squid in tomato sauce, delicious but nothing our Yankee palates had ever had before.

In Cordoba the truth of the expression, "Africa begins in Spain," became very evident. Cordoba could be a town in North Africa, Morocco perhaps. That's not surprising, for the Arabs, when they spread out from Arabia in the seventh century, conquered almost all the lands around the Mediterranean and half of Spain. In those days the Arabs were called the Moors, and their influence is still pervasive today, in the architecture, food, and customs. We walked the narrow white streets of Cordoba as we ate the sticky sweet candies beloved by the ancient Moors and watched the women, their black hair covered by blacklace mantillas, a reminder of the days when, under the Moorish rule, women kept their faces covered. We entered the Cordoba Mosque, which resembled those in Cairo, for it was built about the same time, in 711. Think of what England looked like in 711—it was in the midst of the Dark Ages!

Cordoba was the center of the civilized world then, with the largest mosque in the Arabian Empire. Inside, the striped red-and-white arches stretch away in innumerable vistas and in the middle, we suddenly came upon the cathedral of Cordoba. When the Moors were evicted from Spain, the Spaniards simply used an existing building. But five hundred years of Christian prayers haven't erased the Moorish spirit. [17]

THE GALAPAGOS ISLANDS

"Welcome to the Galapagos!" Phillipe, the naturalist guide who would be with us during our stay on the islands, went off to hunt for our luggage while we looked about us. To be truthful, our spirits drooped. We'd been warned that the Galapagos Islands were not lush but dry. As we looked at the scraggly prickly pear and some sort of gray leafless trees and the dusty red volcanic earth, we decided that forbidding was a better term. We had a sinking feeling that we'd come thousands of miles only to be disappointed, that we'd see few of the fabled animals and birds.

How wrong we were! At our first meal, near the sea, four marine iguanas crawled out of the surf to sprawl by our hastily drawn-up feet, while a great blue heron calmly stood at our elbow eyeing our plates. Next day, while we were swimming at a distant beach, young sea lions played around us like children at recess, and a Galapagos penguin swam up to peer into our snorkle masks. It was evident that these animals considered us one of them; we were iguanas or sea lions or penguins, and none was afraid.

Why weren't they fleeing at the sight of us? The Galapagos Archipelago emerged from the sea bottom as volcanos a million or so years ago, but man has only known of them for a mere three hundred years. The animals and birds that floated or swam or flew from the mainland have evolved unafraid of humans to a degree that never fails to astonish the visitor. More subtle is the symbiotic and adaptive behavior exhibited by the animals and birds. A lizard finding no food learns to dive into the sea and eat seaweed and drink saltwater. A finch on a very dry island pecks a hole in a booby's skin and drinks its blood. Cacti on some islands have developed trunks like those of pines for protection from the giant land tortoises that feed on cactus pads.

One afternoon we came upon a mustard-colored four-foot land iguana chewing a tasty cactus pad. Just then a finch landed nearby, whereupon the iguana raised itself up on all four legs, a signal to the finch that he needed deticking. The finch obligingly hopped onto the reptile's back and proceeded to tweak out the bothersome ticks embedded in the skin around the animal's neck—at the same time enjoying a good meal.

As we sailed among the islands in our small Galapagos fishing boat, Phillipe pointed out that different islands were inhabited by different species of the same animal or bird or plant. We began to realize after the first few days that the endemic species found on the islands are all reptiles, birds, or sea mammals. Land mammals could never have survived the six-hundred-mile journey on their own from the mainland. The wild dogs, goats, and pigs that have created such havoc on certain islands by destroying the habitat and the native species were all introduced by buccaneers, whalers, and settlers.

Today the Ecuadorian government has designated 90 percent of the Galapagos Archipelago as a national park, while the Darwin Research Station is endeavoring to save the most endangered species. We were part of the increased number of visitors allowed by the government to see the islands, all of whom must be accompanied by a guide. We can only hope that such enlightened administration will be continued, for on these fabled islands the mysterious aura that nature ordinarily throws over her works is cast away, and the interaction among man, plant, and animal can be seen clearly. The strange landscape now made sense to us, and we realized that the Galapagos are truly an innocent paradise. [18]

PERUVIAN TEXTILES

American civilization, it has generally been supposed, is much younger than that of the Near East and Egypt. I'm speaking of ancient civilizations, of course. The Mayans of Central America, who flourished about seventeen hundred years ago, A.D., 200, have been thought of as our earliest advanced culture. Now in a desolate canyon in the high Andes of northern Peru, the secrets of one of America's oldest civilizations are coming to light. Five thousand years ago, the ancient people who lived there built elaborate temples and tombs, while artists worked in bone and shell, basketry, and especially textiles.

The arid climate at the ruins of La Galgada has preserved a rich collection of material made in the period between 3000 and 2000 B.C. Because they did not have pottery, their civilization has been termed preceramic. Preceramic Peruvians lived in densely settled towns with massively large temples and elaborate tombs for the upper classes, much as the Egyptians were doing at about the same time.

One of the most intriguing things found in the ruins of La Galgada was a ceremonial firepit chamber, whose floor was covered with white, orange, red, and green feather down. A deer antler was discovered lying on the floor near the firepit. To the archaeologists who uncovered this chamber it was highly reminiscent of the present-day Pueblo kivas of the American Southwest.

Peru is famous for its ancient textiles, and the La Galgada textiles provide important evidence of very early but extremely sophisticated artistic achievement. Wool doesn't seem to have been used, but cotton only, which was probably grown locally. The original colors are well preserved — red, yellow, blue, and black, as well as the natural brown and white. The first woven textiles at La Galgada were narrow cotton belts, showing that the earliest looms were narrow. Bone hairpins inlaid with shell and turquoise were found in a small woven bag along with ornamented discs engraved with birds' heads, their eyes of green or red stones. These were laid beside the body in hope that the jewelry could be used in the next world. [19]

INCA EMPIRE

We had read about the fabulous Incan Empire before we went to Peru, how it stretched from Chile, thousands of miles north to Ecuador. It was so well run that no one went hungry or suffered from unemployment. Areas that are now so dry that only a few primitive farms can produce anything, were once well-watered with a marvelous system of irrigation ditches. Dry-wall terraces were constructed right up to the tops of steep slopes where the Incans grew three hundred kinds of potatoes and as many varieties of corn to feed a large population. Communication was so well organized that those in Cuzco three hundred miles from the seacoast could enjoy fresh, live fish brought by relay runners in waterproof baskets.

It was a fabulously wealthy country, where no one thought of gold as money, only as a beautiful metal. The walls of the palaces and the temples of the capital city, Cuzco, were covered with sheets of gold and silver, the streets were paved with gold paving stones, and in the Temple of the Sun was a garden in which everything was made of gold. Only the stonework now survives in Cuzco, but nowhere in the world are walls more beautifully formed, all done with stone tools. The Spanish and modern buildings superimposed on the Inca foundations collapse with every earthquake, but the Incan stone work is as firm as the day it was laid.

We camped on what was once a broad Incan terrace,

the empty houses with their characteristic slanting window openings looking down on us in our modern tents. It is hard to imagine that this narrow valley supported a large population, but the remains of the agricultural terraces climb to the top of a ridge a thousand feet above us. Now only a primitive farmer and his family raise a few potatoes among the ruins of what was once a thriving community. All day we hiked along the Inca trail in dry eighty-degree heat, then watched the sun going down behind a snowcapped peak at the head of the valley, the night air cooling fast. A few last rays lit the ancient stone buildings above us while smoke rose from the cooking fire of our Indian packers. [20]

QUECHUA INDIANS

A condor soared over the place where we were camped below a fourteen-thousand-foot pass. One of our party was flying a kite and the bird must have come to investigate the strange red-and-yellow intruder. The condor looked like a Boeing 747 as he flew over us, so close that we could see distinctly the upturned feathers at the tip of his ten-foot wingspan. The condors aren't in trouble in the high Andes, where the terrain is so unbelievably rugged that it will never be developed for condominiums or ski resorts.

We have hiked now for three days, and at last at this altitude of eleven thousand feet are beyond where even the Indians can farm. At the beginning of the trek in the broad Urubamba valley, the farms were beautifully fertile and prosperous, an agricultural paradise. As we followed the trail when it turned up ever smaller streams, the farms grew more and more primitive: stone houses thatched with grass, a pig or two rooting about among a flock of chickens, and some cows grazing on dry grass. This is winter in Peru, the crops are in and being taken to barter by one of the only two ways possible, on pack horse or on a man's back. The nearest center at which to barter potatoes for corn is maybe two days away. The only fertilizer used is obtained by burning the dry stalks of the potato vines, then turning the ash into the soil with a crude mattock.

We say "*Buenas dias*" as we walk by the farm yards, and the Indians reply politely, but beyond that they know no more Spanish than we do, for their language is Quechuan, that of the ancient Incas. The women wear a version of the colonial Spanish dress, voluminous petticoats and skirts, even the tinest girls. When the women go off for town they jam a felt hat over their black braids. The men wear those wonderful Peruvian caps with earflaps, sandals on their feet, and enormous bundles tied on their backs in colorful ponchos. [21]

LLAMAS

We associate camels with the Near East, but we saw a flock of American cameloids the other day, a herd of llamas (pronounced *yamas*) being driven across a field with sacks of potatoes on their backs. They were a strange sight with their long necks and little pricked-up ears. They have a camel face, a split upper lip, and two toes on each foot, but they have no hump. We're told that they have a very nasty disposition as does a desert camel and refuse to be ridden, or driven as a draft animal. But they're useful as carriers of produce if the load isn't too heavy, about ninety pounds. They have other uses too, for their shorn wool, while short and coarse, is used to make blankets and sacks, and their dried droppings make good fuel in the high pastures where wood is scarce. The ancient Incas used to dry their meat, which they called *charqui*, from which our modern term of "jerky" comes.

Llamas are one of four American cameloids, the alpacas, vicuñas, and guanacos being the other three. Like the llama, the alpaca has been domesticated, but it is kept only for its fine silky wool from which the finest ponchos are made. The vicuña was never domesticated, and now its numbers have been badly depleted by overhunting. I'm told that the Peruvian government now has forbidden their killing, but it is still too soon to know if they will survive. The fourth cameloid is the guanaco, few in number, from whom the llama, vicuña, and alpaca are descended.

In the days of the Incan Empire, animal management was handled better. Each province was divided into four quarters, and each quarter held a community hunt every four years, the emperor joining in the affair with his subjects, for wild animals were considered state, not royal property. These ancient civilizations were always careful to leave enough animals to perpetuate the breed. Perhaps we have something to learn from the Incas. [22]

PERUVIAN FOOD

One of the best things about traveling is to be able to sample food different from that at home. And food in Peru is different! Take their potatoes, for instance. In the native markets we lost count of the different kinds, white potatoes, yellow potatoes, purple potatoes, all shades in between, and every imaginable size and shape. We can attest to the variety of tastes and textures and the imaginative ways of serving them, too. Not for the Peruvians the ubiquitous boiled, mashed, baked, and fried.

We were told by our guide that the Incas were the first to develop potatoes, which were unknown in Europe until they were brought back by the first explorers. Eventually the Incas grew as many as three hundred varieties. They were the first, too, to use freeze-dried potatoes. They would expose sliced potatoes to freezing, thaw them and squeeze out the moisture, then dry them. Corn is another development of the Western Hemisphere, and the Incas grew enough varieties to make the head of a modern plant geneticist spin. We couldn't begin to keep track of the varieties we saw. Ancient Incas made a beer from a purple variety of corn, called *chicha*, still available today. When we tried some after a hard day's hike, we found it good, its purple color notwithstanding.

In order to support a large population in the mountainous country beloved of the Incas, the early civilization built terraces on the steep hillsides to prevent erosion. A few are still used today, but many are just faint outlines, stepping up the mountainsides until one's head swims just to follow them. The effort that went into making them is unimaginable to modern minds. What danger was encountered building the stone retaining walls over sheer cliffs, then bringing in baskets of clay, gravel, and soil from the lowlands miles away. All this effort supposes a mighty organization, and the Incan Empire was one of the most efficiently run governments ever known.

Why then was it overthrown so easily by the Spaniards? One reason was that the Europeans had already infected the Indians with measles and smallpox. Also, there was civil war raging in the empire, and all the Spaniards had to do was to wait until one side betrayed the other. The Indians must regret their actions still, for their lot under the Spaniards and their successors has never come close to what it was under the old empire. [23]

PERU

Who hasn't been intrigued by the fabulous lost city of the Incas, Machu Picchu? The pictures show it perched on a ridge whose sides drop down a thousand feet to a river. For four hundred years it remained a secret of the Indians until in 1911 American Hiram Bingham scrambled somehow up the snake-infested cliffs to find a deserted city overgrown by the jungle. Who built it and why, no one really knows, but today, cleared of the overgrowth, it has become one of the wondrous places in the world and visited by thousands.

We decided we must see it, but instead of taking a train and then a bus to the city, we'd spend six days hiking on the Inca Trail that winds through the Andes. This way we'd see the other remarkable towns and forts that guarded the access to Machu Picchu. On looking at maps we realized how little we, along with other North Americans, knew about the geography of South America. For instance, Lima, the capital of Peru, although on the west coast of the continent, is almost directly south of Charleston, South Carolina, on the east coast of North America. Therefore, we'd be in the same time zone except that Peru keeps standard time. And because we'd be below the equator, we'd be there in Peru's wintertime, which in Lima means fall clothes because its nearness to the Pacific tempers the cold. While in the high Andes it would be a different story: we needed to be prepared with winter underwear, gloves, and down jackets. At the same time, the high altitude would make the sun fierce during the day.

The rivers on the Pacific side of the Andes are very short. Once they leave the mountains, they flow through desert country, where the narrow valleys may be extremely fertile. The mountains crowd close to the Pacific, and in a half hour's plane flight from Lima on the coast, you've crossed over into the Atlantic watershed and into the Amazon drainage basin, which stretches three thousand miles to the sea. Peru has unnumbered snowcapped peaks over twenty thousand feet, some still unnamed and many unclimbed. We've been warned about the effects of high altitude, about not drinking glacier water, about pickpockets in Lima. We've waterproofed our hiking boots, packed dark glasses for the bright sun, resurrected winter underwear against the freezing night temperatures, and we're ready to start hiking the Inca Trail to Machu Picchu.   [24]

MOUNTAIN TRAILS

We have been following the Inca Trail to Machu Picchu the past few days. When it was built, it was a six-foot-wide paved highway, with steep steps carrying it over precipitous slopes. By now, much of the trail on the mountainsides has been obliterated by landslides, and we must follow a dirt path made by the local Indians. Where the trail still exists, its width has been narrowed to about a foot and a half, and occasionally even that has been wiped out. In a very gingerly manner we make our way across three or four logs placed across the gap, trying not to look down into the three-thousand-foot abyss.

We know that the Incas maintained a magnificent system of highways that stretched for thousands of miles over deserts and mountains and down to the jungles. On the way, there were well furnished rest houses for the traveler, forts and lookouts, and large towns perched on mountainsides. We have camped in the midst of a number of these spots. One, Llactapata, was the bureaucratic center for the region, the sort of place where the Incan version of the Internal Revenue Service kept its records. Stepping up a steep hillside, the houses need only roofs to make them habitable again. Another town, discovered only in the 1940s, Winywayna, by a magnificent waterfall whose stream flows into the mighty Urubamba, is placed on a site so impossibly steep that you wonder it hasn't slid down into the canyon. But of course the Incas built so well that four hundred years of earthquakes haven't dislodged a stone.

At the top was the usual rounded temple to the sun, just below a series of overgrown terraces, and a staircase of one hundred eighty steep steps led down to the dwelling quarters. The town was watered by an ingenious series of irrigation tanks and troughs brought somehow around the ravine from the waterfall. With a great sense of triumph, one of our Indians with the aid of some long bamboo spikes was able to get the topmost water tank working. Down in the lower town there was a maze of houses and courtyards all perched on a narrow platform. In the farthermost courtyard was a large portal, which led out to nothing. Was it a place from which to worship the waterfall and the river three thousand feet below? Or did a priest stand in the doorway holding a sacrifice to the condor which would swoop down and carry it away? Who knows? [25]

DEAD WOMAN PASS

    The tents were already up and waiting for us on the other side of Warmiwaunca, or, as it is known in English, Dead Woman Pass. We had started off in the pouring rain in the morning, but as we gained a little altitude the rain turned to snow. The trail wound up and up without let-up in order to gain the fourteen-thousand-foot altitude of the pass. We found it better not to look up to see where we were going, but to simply fix our eyes on the boots of the person in front and to concentrate on keeping a steady pace and breathing as evenly as possible in order not to waste energy. We had no breath to use on conversation, and were warned that we mustn't stop our steady pace until we reached the top.
    In an hour and a half the trail suddenly pitched over the other side. We had made it! The wind was blowing hard as we crouched behind rocks and sucked on some energizing hard candies. Just then our Indian bearers came up the trail with their huge packs containing our food and tents, bare feet in sandals as they sloshed through the snow. They weren't chewing candy, but each had a wad of cocoa tea stashed in his cheek, with a bit of limestone added. This is supposed to alleviate the effects of cold and hunger. They too stopped for a rest. One of them got out

his flute made of bamboo, and, sitting in the snow, played one of the plaintive, melancholy tunes, heard often in the Andes. Then they squelched down the other side of the pass, and when we came into camp hours later, they had our tents all up and huge pots of hot sugary tea waiting for us.

We spread out our sleeping bags, and by the time dinner was ready, it was pitch-dark at six. In the Andean winter, night comes on early and quickly. There is no sitting around a roaring campfire; at this altitude wood is scarce, just enough for the Indians to cook up their pots of potatoes. Then they huddle together under a blanket while we, flashlights in hand, upzip our tents. It is cold tonight. Boots and heavy woolen pants come off, but long johns, woolen caps, and socks will accompany us to bed. As we drift between sleep and awakeness, the Indian pipes us to sleep with his flute. [26]

TREK TO MACHU PICCHU

The great moment of our trek came when we hiked around a shoulder of a mountain, and there, far below and far away, were the fabled ruins of the sacred city of Machu Picchu. Then we lost sight of it as the trail swung away in order to lose altitude. The next day we saw it again from the ancient Gateway of the Sun. Much closer now (we could pick out individual buildings), it lay on its green saddle between two mountainous towers that were in turn dwarfed by the mountains rising from the stupendous canyon of the Urubamba River.

This was the way the Incas themselves first saw the city, after their feet had carried them over the same mountain passes we had crossed. Now the tourists come by train from Cuzco, then up the switchback road by bus to the inn and modern gate. They spend a few hours and at three they have gone again, back to Cuzco. After camping out for seven days, we enjoyed the comforts of the inn, but were up early in the morning, watching the mists from the river below clear away and peak beyond peak glow in the rising sun.

We had a most knowledgeable guide for two days and then wandered by ourselves through the ruins. Here was the place where a puma was chained to guard a sacred woman; this round temple was sheathed in gold, and this low terrace was where experimental plants were grown, where they were safe from the chill Andean winds. In the middle of the city, one feels safe and secure, but wander off to its edges and one's head swims looking down into the abyss.

Why was a city placed in this extraordinary place so remote that it was lost sight of by the white man for four hundred years? What was its purpose? No one knows for sure. Perhaps that is what adds to its sense of mystery. All we know is that it is a very special place.

The final night three of us were fortunate to go with an Indian guardian into the darkened ruins, an expedition frowned upon by the authorities. We had to be very quiet, and indeed we were too overawed to speak. Holding hands, we came into the central plaza, where we stood silently looking into the brilliant sky apparently so close to us. The Southern Cross swung overhead, and it seemed profoundly right that the Indian should point out a stone shaped just like the constellation, positioned so that it pointed directly at it. Surely the spirits of the Incas were there with us. [27]

CUZCO

We are back in Cuzco, our trip in Peru almost at an end. Cuzco is a fascinating city, the old capital of the Inca Empire. The air at eleven thousand feet is thin and cold and has a delightful spicy odor. The city nestles in a valley whose entrance is guarded by a snowcapped peak. It's one of low red-tiled roofs, broad plazas, and imposing seventeenth century churches. The Spanish followed the plan of the Incan city, pulling down the temples and houses to build their own. But enough foundation walls remain to give an idea of the original magnificence. Ironically it is the Spanish and more modern buildings that collapse in earthquakes.

Gringos wearing ponchos rub shoulders with Indian women spinning the wool with which to weave the ponchos, while Indian men sit on sidewalks and play on their flutes their melancholy Andean airs. Their ancestors' world came crashing down when the Spanish arrived seeking the gold that the Incas regarded as a beautiful metal, not as a medium of exchange. Fortunately for Pizzaro, an empire wracked by civil war and already devasted by the white man's diseases fell easily into his hands. How he dealt with the emperor, promising him freedom when his prison cell was filled with gold, then treacherously cutting his throat, is a story that is still alive. The history of the white man's treatment of the natives thereafter is not a pleasant one. Millions were enslaved and treated worse than the Spaniards' animals; those who could fled into the remotest valleys. Fifty years after the Spanish arrived in 1533, the last Incan, Tupu Amara, who had managed to maintain a shrunken kingdom in the fastness of the Andes, was betrayed and brought in chains to Cuzco. There his people crowded to the rooftops and set up a great wailing. The emperor raised his arm and the people fell silent, as he cried, "Honor your mothers and remember your great heritage and be quiet for my death." He was dispatched with the utmost brutality.

Today the Indian lives a life of unceasing toil and poverty. Those who leave their primitive farms for the city find conditions that we northerners can't conceive of, so terrible are they. It's hard not to conclude that the Indians were better off under the rule of the Incas. [28]

Planets & Space

A SHORT HISTORY OF ALL HISTORY

About twenty billion years ago a diffuse cloud of gas, some ten trillion miles in radius, began slowly to coalesce, pulled together, it is theorized, by the force of its own gravity. With the passage of time the atoms of gas grew closer, the attraction of the gravitational forces grew stronger, and the cloud began to contract even more, so that after some ten billion years it was no more than 864,000 miles across. At this point, the temperature within the central region of the cloud was high enough to ignite nuclear reactions. Hydrogen was fused to helium, energy was released, and a star was born.

During this extended period of creation, lesser condensations were taking place in the cooler region surrounding the great cloud of gas. As a result, decidedly smaller and cooler celestial bodies were created. Five billion years ago, there were nine of these bodies circling the central star. One of the circling bodies contained on its surface large collections of hydrogen and oxygen bonded together in various densities. There were vast concentrations of this liquid substance and in this warm chemical sea, molecules began to mix and work upon each other, binding together until, at some point, some of the more complex of these molecules were able to replicate themselves. About four and a half billion years after the first replicating molecules appeared, one of the species that resulted from this curious process began to apply names to things. It called one of the smaller, cooler clouds of the coalesced gas, Earth. The larger, brighter sphere, which was determined to be some ninety-three million miles from the mother planet, it called Sun.

The new species produced many ingenious, lovely things in its time. It also generated abstract concepts and theories. One theory holds that, in due time, the sun and all its planets will cease to be. According to this hypothesis, a star the size of the sun has a life span of about ten billion years, which means that Earth's star is well into middle age. Some five billion years from now, the sun will begin to swell into a vast sphere of heated gas which will vaporize all of its children, the planets. After approximately one billion years in this state, the sun will use up its hydrogen fuel and begin to collapse into a dense white-hot star. Slowly, in the remaining millenia of its life, the sun will radiate the last of its heat into the empty universe and fade into cold blackness, its presence extinguished forever. [1]

THE WINTER SKY

Skygazing can be at its best on clear winter nights. Moisture content is low in cold air, enabling light from the moon, planets, and stars to reach us undistorted and undimmed compared to other times of the year. Without visual aid, an observer in an area free of obstructions can see the moon, up to four planets, and approximately two thousand stars on a dark, cloudless night.

Although it appears to be the largest object in the nighttime sky, actually the moon is the smallest. This illusion is created by relative distances, as the moon is only

238,000 miles from the earth compared to the closest planet, Venus, at fifty million miles away, and the closest star (aside from the sun) at more than twenty-five trillion miles distant.

Venus, Mars, Jupiter, and Saturn can all be seen by the unaided eye at various times in February. Because of their movements around the sun, planets change their positions relative to the stars. These changes may be perceived weekly for Venus and Mars, much more slowly for Jupiter and Saturn. Stars are also moving, sometimes at great speeds. Their movement is not noticed because of the astronomical distances that separate them from the earth. Given enough time, these movements would become discernible. In fifty thousand years, for example, the Big Dipper will not be recognizable.

Planets do not produce their own light. They shine with light reflected from the sun. This light seems more steady to us than the often-flickering light that we see from stars. This is again because of the vast distances that the light produced by stars must travel compared to that for the planets.

Venus shines brighter than any star, and is only seen for a few hours either after sunset or before sunrise. In February, Venus is visible in the western sky from shortly after sunset until approximately nine o'clock. It will be the first "star" to appear each night in February.

Mars, which has a reddish hue, and Jupiter, which shines as brightly as Venus, follow similar paths during February. They rise in the east in the evening and are visible throughout the night, as is Saturn, which is about as bright as the brightest star (Sirius). Looking through a strong set of binoculars, an observer may be able to see up to four moons circling Jupiter, as well as Saturn with its rings. [2]

ARABIAN ASTRONOMY

In the Arabian desert, travelers navigate as sailors do at sea. The terrain has long since been robbed of its landmarks by thousands of years of wind erosion. Only vast featureless plains remain, interrupted by seas of dune sand. One sandstorm may change the look of a familiar dune field; the next fierce wind rearranges it again. Because travel was a way of life for the bedouin, these nomads had to know the sky as we know our city streets. To find their way, the Arabians mastered the sky long before history was recorded.

Arabian astronomy reached a significant phase with the translations of the ancient Greek philosophers, carried out under Harun Al-Rashid (765-809), the hero of The Thousand and One Nights. His father, Al-Mansur, had been impressed by a book from India that dealt with the stars and eclipse predictions. He had the *Brahma-Sphuts-Siddhanta*, or the *Opening of the Universe*, translated into Arabic and the book remained the standard work on astronomy for half a century. Following his father's example, the new Caliph ordered Ptolemy's *Megiste Syntaxis*, or *Great System*, translated into Arabic. The resulting work, the *Al-Magest*, became the major source work for Arabian astronomy, and Ptolemy's work was preserved for posterity.

In turn, Harun's son and successor, Al-Mamun (786-833) initiated the next phase. He erected the first observatory of its kind in Baghdad to verify the fundamental elements of Ptolemy's *Al-Magest* and perform additional observations. This encouragement of astronomy came at a time when science was essentially non-existent in Europe. From a practical standpoint, the caliphs wanted to refine the lunar calendar and make better predictions of lunar eclipses and other celestial events. More significantly, the religion of Islam emphasized the study of the vast heavens and the order in the universe as signs of the greatness of God. In one passage, the Quran, the holy book of Islam, states: "In the creation of the heavens and of the Earth, and in the succession of the night and day, are marvels and signs for those of understanding heart."

In the Baghdad Observatory, astronomers performed a delicate geodetic measurement of the length of a terrestrial degree. They aimed to determine the size of the Earth. Their results were remarkably accurate: circumference 32,830 kilometers, diameter 10,460 kilometers. Among those who took part in this measurement was Mohamed Al-Khwarizmi (780-850), who also compiled the oldest astronomical tables. After being translated into Latin in 1126, the tables became the basis for other works in both the East and West. Al-Khwarizmi also composed the oldest work on arithmetic and coined the term "algebra." His own name, converted into "algorithm," denotes the system of numerals widely in use today. [3]

LIGHTNING

Lightning is the leading cause of fires on farms, as reported by the National Board of Fire Underwriters (now a part of the American Insurance Association). It starts thirty-seven percent of the fires resulting from known causes. Of even greater significance is the fact that every year in the United States four hundred people are killed by lightning and over a thousand injured. There is also a tremendous unlisted amount of property damage without any fires.

Lightning occurs over most sections of the country with varying frequency; New York State can expect about thirty such days per year, generally in the late spring and summer, usually on hot humid afternoons and in the early evening.

Lightning, like all electricity, generally follows the shortest, easiest path to the ground; however, because of the tremendous voltage that can be expended in a lightning discharge, it often follows simultaneously a series of paths in the same direction.

It is estimated that a lightning stroke can start as hundreds of millions of volts potential between cloud and earth, and still discharge millions of volts at the point of contact — the building, object, or ground point that it hits. The current duration of the stroke is short, but because of its tremendous strength, averaging from 10,000 to 218,000 amperes, it can generate tremendous heat if the conductor carrying it away is insufficient in size.

If lightning hits dry wood, the great pressures around its path will shatter the wood. Sap in a tree will be instantly turned to high-pressure steam causing the tree to explode, and any highly combustible material like hay or straw along the bolt's path may be heated to ignition temperature. Part of the reason for such an explosive action is that electricity flowing through a poor electrical conductor (such as green wood) creates far more heat than when flowing through a good conductor (such as copper or aluminum).

The path of the lightning stroke is not completed when it hits the top of a building or lightning rod, but it must continue on until it reaches earth or an adequate conductor to earth. It is extremely dangerous to provide an easy path for a buildup of static energy on the top of an earth object and not provide a large enough path to discharge the stroke should it occur.

Every year thousands of animals are killed while standing against or near ungrounded fences that are struck by lightning, or charged because they are attached to or are near a tree hit by lightning. It is estimated that lightning can follow a metal fence for a mile or more. If a fence is attached to a building or near one, it can conduct the lightning discharge to the building where it can cause damage and death. Thus, fences need to be grounded in a manner similar to other structures. [4]

SPACE TRASH

On a clear night sometimes you can see a starlike object winking its way across the sky. If you look up with binoculars just after sunset, you might see four or five of these twinkling specks of light passing every hour. A few of them are artificial satellites, evidence of man's increasingly productive use of space. But most of them are garbage.

"There are only about 235 operational payloads now in orbit," explains Donald J. Kessler, an astrophysicist at the Johnson Space Center in Houston. The rest of the estimated 10,000 to 15,000 bodies circling the earth are pieces of useless junk, which according to Kessler could soon pose a greater hazard to spacecraft than meteoroids.

Space junk ranges from nonfunctioning satellites and spent rockets to tiny fragments left over from collisions and explosions of spacecraft. "There have been about fifty explosions in space that we know about," Kessler explains, "and those have generated about 60 percent of the total tracked population of junk."

The debris is thickest and collisions are most likely in the region from 100 to 1,200 miles above the ground, the area of greatest unmanned activity to date. Kessler says the most probable accident is the ramming of two inactive objects, such as an old rocket body and an explosion fragment. That collision would unleash a rain of millions of new pieces, which in turn could hit other spacecraft. Such collisions, rather than explosions, will probably create most of the space junk of the future.

For all the mess up there, the risk of collision for any particular craft is low. If the space shuttle were in orbit for a full year, for example, there would be less than one chance in ten thousand of its being struck by a known piece of junk. Still, the number of stray items is growing by 11 percent a year, and scientists expect that something will crash into at least one hapless spacecraft within the next decade. [5]

METEORS AND MEETEORITES

People often confuse the terms meteor and meteorite. The word *meteor* is derived from the Greek *meta* (beyond) and *eoros* (a hovering in the air). Meteors are solid objects in space that come into contact with the earth's atmosphere and burn up because of friction. Meteorites are similar to meteors, except that meteorites actually reach the surface of the earth. They land because of a number of reasons, mainly size, composition, speed, and angle of the object when it collides with the atmosphere. The term "shooting star," which is often applied to both meteors and meteorites, is a misnomer, as these objects have no direct connection with stars. In fact, meteors and meteorites are composed mostly of debris from comets and asteroids. They range in size from microscopic particles to huge boulders weighing fifty thousand tons or more.

A meteor shower usually is caused by the interaction of the earth's atmosphere with debris from a decomposing comet. This happens when a comet passes through the path of the earth's orbit around the sun. Heat from the sun causes the comet, which is composed mainly of rock and frozen gases, to melt and release particles into space. These particles remain in the comet's path. When the earth crosses this path, and our atmosphere comes into contact with these particles, friction causes them and the surrounding air to heat up, producing the bright, sometimes colorful streaks in the sky. Often, the atomic composition of particles can be determined by the color of the meteor's trail. Many meteor showers are regular, annual events. The Geminid in December and Perseid in August are the two most spectacular annual meteor showers.

Meteors can be seen in fewer numbers by a patient observer on almost any clear night (they do occur during the day, but cannot be seen because of the light from the sun).

It is estimated that as many as 100 million meteors and countless billions of tiny micrometeorites collide with the earth's atmosphere daily. Larger meteorites, such as the one that created the Barringer Crater in Arizona (275 miles in diameter), are thought to collide with the earth approximately every ten thousand years. Atmospheric, geologic, and vegetative forces have prevented the earth from resembling the pockmarked appearance of the moon. An estimated four million tons of space matter falls to the earth annually. [6]

ASTEROIDS

Hurtling and tumbling through space, mainly in the zone between the orbits of Mars and Jupiter, are thousands of rocky and metallic and organic little worlds called asteroids. The first asteroid was discovered at the very beginning of the 19th century and the rate of discovery of new ones has increased steadily since then. But no one has ever studied an asteroid close up. There are, as yet, no images of their surfaces. The best we have done so far is to photograph Phobos and Deimos around Mars; the outer small satellites of Jupiter; and Phoebe, the outermost Saturnian moon. These are small worlds between a few kilometers and a few tens of kilometers across, which, largely because of their peculiar orbits, are thought to be asteroids captured when they wandered too close to larger nearby planets. Some asteroids on occasion come close to the Earth, but, so far as we know, there are no asteroids in orbit about the Earth.

Although we have never scrutinized a bona fide asteroid from short range, we may nevertheless know a great deal about some of them — because most meteorites that fall on the Earth probably originate from asteroids. They are pieces chipped off by collisions with space debris —

generally a comet or another asteroid—which, over immense periods of cosmic time, are eventually swept up by the Earth. The trouble is that we do not know which meteorites come from which asteroids. There are many varieties of each. Some rare meteorites have apparently been ejected from the Moon or Mars. Many are still of unknown origin. The spectral properties of iron meteorites are similar to those of M-type (metallic) asteroids; perhaps iron meteorites come preferentially from M-type asteroids. C-type (carbonaceous) asteroids are very dark; they reflect only a few percent of the light that falls on them. The same is true of the organic-rich meteorites called carbonaceous chondrites. Could the C-type asteroids (and the moons of Mars, which resemble them) be worlds composed significantly of ancient organic matter from the early history of the solar system?

The largest asteroid known, 1 Ceres, 1,000 kilometers (about 600 miles) in diameter, is of this very dark, slightly red variety. Such asteroids, all presumably containing abundant organic (carbon-rich) matter, are very common; about sixty percent of the main-belt asteroids between the orbits of Mars and Jupiter are carbonaceous. [7]

GETTING READY FOR HALLEY

"We've been looking for the last three years and we intend to look again in the coming season," said Michael Belton, referring to the current search for Halley's comet. An expert on planets and comets at Kitt Peak National Observatory in Tucson, Arizona. Belton is leading an attempt to make the first sighting of Halley's comet before its grand sweep past the earth and the sun in 1985-1986.

Halley's comet has passed this way before; its last trip was in 1910. There are several scientific reasons for trying to make the earliest possible sighting of this previously known comet on its upcoming trip; among them are to determine its physical state while far from the sun, to measuring the size of the nucleus before it starts shedding material, to gain increased knowledge of the comet's orbit as a navigation aid for planned cometary space missions, and to assist in pointing the space telescope correctly. One nonscientific reason is that whoever is first to sight the comet will be assured of a place in astronomical history. The most famous comet observation of all time was made by an eighteenth-century German farmer, Johann Palitzsch, who was first to spot Halley on its trip through the inner solar system in 1758. Without accomplishing that feat he would surely have remained unknown. Instead, Palitzsch is remembered in any number of historical accounts and books about comets.

Palitzsch upstaged the professional astronomers of 1758 by sighting Halley's comet on Christmas Day, on its first return following the prediction by the English astronomer Edmond Halley. The return not only verified Halley's prediction but also lent increased weight to Isaac Newton's theory of gravity and gravitational effects on the motions of objects in space. Needless to say, the return had been eagerly awaited by the scientists of the world—so much so that Voltaire is said to have quipped that "astronomers never went to bed in 1758 for fear of missing the comet." [8]

MARSQUAKE

When space voyager Mariner 9 flew at an altitude of two thousand kilometers above the planet Mars, its cameras took a series of photographs that fitted into a mosaic of pictures showing a portion of a great complex of canyons named *Valles Marineris*. The photographs were taken on July 3, 1976, of a side canyon called *Ganghis Chasma* situated about 1,900 kilometers south of where Lander I touched down three weeks later. The picture looks almost like the Great American Desert, as parts of Colorado, Arizona, and New Mexico used to be called. But no, the canyon was on a planet millions of miles above us.

Why did scientists sit up and take notice? Most of the photographs showed a smooth plain, but in the lower two-thirds the surface seemed to have been eaten away from beneath, collapsing into chaotic depressions. This process, called *sapping*, is caused by the removal of underground water or ice. The widespread effects of sapping have altered and eroded the canyons, which originally were formed by tectonic activity, or, in other words, the motion of faults in a gigantic earthquake, or "Marsquake," long ago, causing the ground to sink. The evidence of sapping here, and in many places has contributed to other evidence of underground ice on Mars, perhaps almost everywhere. Ground-based radar and spectroscopic data have been interpreted to indicate the presence of liquid water near the surface in the region called *Solis Lacus* (Lake of the Sun). These interpretations are controversial, however, as they appear to conflict with data from the water vapor measurements of the Viking orbiters. [9]

SURVIVAL IN SPACE

Unprotected in the vacuum of space, a human being would survive for only a few minutes before swelling up (because water vaporizes) and becoming freeze-dried. Accordingly, without artificial life-support and protective systems that emulate the earth's environment, manned space flight would not be possible.

Humans evolved in an atmosphere composed primarily of oxygen and nitrogen at a pressure of approximately fifteen pounds per square inch. Although it may often seem very hot or very cold, the earth's atmosphere provides a relatively benign temperature environment within an average range of about plus or minus one hundred degrees Fahrenheit. It seldom gets colder on Earth than minus eighty degrees or hotter than one hundred thirty degrees. By contrast, space lacks a true atmosphere, and the temperature may range from minus four hundred sixty degrees, an approximation of absolute zero, to three hundred degrees or higher.

The main requirements for survival in space are oxygen at an appropriate pressure, supplies of food and water, and systems to control humidity, ensure a comfortable temperature range, remove carbon dioxide and other gases, such as sulfur dioxide, and handle our solid and liquid wastes. Designing equipment to satisfy these requirements within the smallest spaceship of all—the space suit—is especially difficult.

The fundamental purpose of the space suit is to establish around the astronaut's body an artificial atmosphere containing oxygen. To maintain life and to saturate the blood with oxygen, the oxygen pressure in that atmosphere must be greater than approximately two and a half pounds per square inch, but there is no need to have a maximum pressure higher than our normal one. The space suit also must protect the astronaut from other hazards, such as bombardment by micrometeoroids. Because the intensity of sunlight is much greater above the shield of the earth's atmosphere, the side of the suit facing the sun may get very hot, while the other side, exposed to deep space, may get very cold. The suit therefore must be insulated to protect the astronaut from these temperature extremes. The earth's atmosphere also guards us from some of the high-energy radiation in space; the space suit accordingly must provide some degree of radiation protection. [10]

AURORA BOREALIS

Frequently, New Englanders who are lucky enough to be about late in the evening witness a spectacular display of the aurora borealis, or nothern lights. The sky is streaked with glowing lavenders, pinks, and greens.

Whenever people share their memorable experiences seeing the northern lights the inevitable question arises — What causes this phenomenon? As I began to search, it became clear that no one has yet devised a thorough explanation for what causes the auroras. But there are some well-accepted theories among astrophysicists.

The sun emits a steady stream of hydrogen into the solar system. This hydrogen is a by-product of the nuclear reactions that are the source of the sun's energy. The cloud of hydrogen particles becomes less dense as this solar wind ranges far out into the solar system. Scientists call this thin, solar atmosphere the *corona*. When it reaches the earth, the corona is only one-one billionth as dense as that of the earth.

When the solar wind approaches, the earth's magnetic field funnels the particles toward the polar region. Here, the gases of the earth's atmosphere and the highly charged solar particles react, and the aurora borealis is born. Because the sun's activity, and thus the strength of the solar wind, is highly variable, the auroras tend to be more frequent and intense when the sun is more active.

How do the auroras glow? A rough analogy could be made using such familiar devices as fluorescent light bulbs and neon signs. In these cases, certain gases fill a glass tube devoid of air, and electrical energy causes the gas particles to glow. Each kind of gas emits a certain color. This may be roughly what happens with the northern lights, as atmospheric gases are charged by the energy of the solar wind.

Somehow, it is satisfying to know that something as beautiful and celestial as the northern lights still has about it a rather nebulous origin — an air of mystery. [11]

THE TECHNICOLOR® SKY

Who has not been moved by the beauty of a magnificent sunset washing the evening sky with flaming hues . . . or by a rainbow arching across the landscape after a passing storm . . . or by the shifting streamers and pulsating glow of an aurora? Such spectacular sky displays of color and light are staged by the interaction of sunlight with the ocean of air surrounding our Earth.

Sunlight is a mixture of all the colors of the rainbow, ranging from the long wavelengths of red to the short wavelengths of blue and violet. In traveling through the Earth's atmosphere, sunlight is scattered by tiny air molecules and by dust particles. The shorter blue waves of light are deflected the most and scattered in all directions, creating the familiar blue daytime sky.

A host of startling but colorful effects are possible when the sun is near the horizon. Not only are light rays scattered more by the denser air near horizon level but the rays also are bent, or refracted, displacing celestial objects upward from their true positions. At sunset, by the time the sun's rays reach our eyes, nearly all the colors have been scattered out by particles in the air except for the longer red and orange wavelengths. Thus, the sun appears red. Clouds and dust in the sky reflect the prevailing color, enhancing the display.

The nearer the sun is to the horizon, the more its image is affected by refraction. The bottom of the solar disc is bent upward more than the top, making it appear flattened or squashed. In addition, air layers of varying densities may impart a rippled look to the sun's edge or sometimes break it up altogether.

If the sun sets over a clear straight horizon, the last bit of sun to vanish sometimes turns green—the so-called green flash. At the horizon, only red and green colors remain on the sun after other colors are absorbed or scattered by the air. Green is bent upward more than red; therefore, green is the last wavelength to disappear. [12]

PROJECT SENTINEL

Project Sentinel, The Planetary Society's major project in the Search for Extraterrestrial Intelligence (SETI), is being greatly expanded. The membership-supported research is already the world's most advanced SETI project currently operating, but the planned expansion will make it sixty-four times more powerful.

The expansion hs been given its own name, META, for Megachannel Extraterrestrial Assay.

META, to be completed this year, will allow the SETI project to observe 8.4 million channels at a time. "META makes the system the biggest analyzer on Earth."

The fundamental problem of SETI is that it is difficult to cover all potential radio channels, all parts of the sky, and all reasonable signal strengths. To accelerate the search, we must try to anticipate the types of signals that might be transmitted. The first assumption most SETI programs make is that aliens have built radio beacons to let other civilizations know of their existence. (Ordinary signals such as Earth's television and radar would probably be too faint to detect with present techniques.)

In Project Sentinel we try to deduce which radio channels they might use. Fortunately, there are certain channels that atoms and molecules of our Milky Way galaxy normally broadcast on—the "magic" frequencies. We suspect that an alien civilization that wanted to make itself known to others would broadcast on or near one of these distinctive channels. Sentinel is currently tuned to the strongest of these, the hydrogen signal at 1420 megahertz. An alien beacon might send an extremely carefully tuned (ultra-narrowband) signal near this frequency, because such a signal would stand out amidst the nearby natural hydrogen radiation and the general galactic noise.

But our troubles are not over even if we have chosen the right magic frequency. The problem is that everything in the universe is moving, and any signal will be changed by the transmitter's motion toward or away from us, a phenomenon known as the Doppler effect. For this reason, a signal transmitted on one radio channel is received on a different one. In Project Sentinel, we assume that the aliens have corrected for the change due to their motion toward or away from our Sun. However, it is conceivable that they might instead decide to shift transmissions to the same channel that signals from the center of the Milky Way would be on, or even to the channel corresponding to the center of the universe, both of which are outside the current range of Sentinel. Project META will allow us to detect them if they have chosen either strategy. [13]

Energy

ENERGY AND AGRICULTURE

Some people are complaining about the use of energy on the farm. Some have proposed that we go back to a "peasant" type of agriculture, including the use of hand tools, as a means of saving energy. They may not be aware that the production of all food and fiber by agriculture requires only 2 to 3 percent of the total energy used in the United States yearly. All the animal industries probably require no more than 1 percent of the total energy used in the country. To eliminate all energy use by agriculture would have very little impact on the total energy picture in America. We should, of course, develop new energy-saving technologies and alternate sources of energy as a means of conserving energy and being as efficient as possible on the farm. But, the answer is not to turn back the clock and destroy the most efficient food-producing system in the world, and along with it our high standard of living.

The total farm population numbers less than 5 percent of the total population. Farmers have a difficult time getting enough help at harvest time. This has forced some to resort to mechanization or to the importation of foreign-labor to harvest their crops. Not many people (even those out of work) are interested in the back-breaking labor on the farm. A recent report by the Council for Agricultural Science and Technology indicated that producing crops by 1918 technology would require about one-third of the total United States work force (eight times the number now involved), plus sixty-one million horses and mules and half the United States cropland, just to feed the horses and mules. This kind of agricultural production would result in considerably higher prices for a smaller amount of food and subsequently a lower standard of living in the United States. Moreover, where would the additional people needed on the farm come from? We could not find them. [1]

SOLAR GREENHOUSE

Solar greenhouse designs reflect more than sunlight. They demonstrate the adaptability of this concept to the particular site and to the individual's personal tastes. There are limits to siting solar structures, but within an optimum range, much can be tailored to the specific circumstances.

The goal of the design is to maximize solar insulation and minimize loss of heat in the cold months. The greenhouse also must be capable of tempering extreme temperature fluctuations. Many facets of design are involved in reaching this goal, one being the angle of glazing used on the south-facing wall of the structure.

As plants are grown in the greenhouse, the most obvious need is for light. Traditional designs for greenhouses have the long axis run north–south with the roof divided into two, of equally shallow slope, with glazing on all walls and roof. This design was created for a European climate of predominately diffuse light (cloudy) in winter. New England has a trend of clear, sunny days when the temperature plummets, so the solar radiation is direct not diffuse when we need it most in the greenhouse.

The angle of glazing is determined partially by the anticipated use of the greenhouse. The maximum amount of radiation is transmitted through glazing when the radiation is perpendicular to the surface. The Earth's tilt on its orbit around the sun is 23½ degrees from vertical, so we have seasonal variations in weather and the sun's path across our sky. The greenhouse is a fixed-angle structure, whose designer or owner–builder must determine for which season the glazing should be optimal.

The general rule of thumb for greenhouses designed for winter use is to add 15 degrees to your latitude for the angle of the glazing. This results in a 60 to 70-degree angle. If the greenhouse is intended primarily to extend the growing season in spring and fall, the 40 to 50 degree angle receives the highest incidence of radiation. Contributing factors, however, may make other angles viable. In a northern climate, the typical snow cover found in winter makes vertical (90 degree) glazing feasible. Snow reflects between 40 and 95 percent of the light striking it, depending on its freshness, which can add 15 to 30 percent to the solar heat output of a vertical collector, depending on the topography.

You can experiment with specific calculations and numbers to provide a basis for comparison, although different studies often provide different results. Many greenhouses are adapted to local situations by using glazing on different angles, such as a 60 degree roof pitch and vertical walls. Combined with insulation, and thermal mass, greenhouses will work and become an energy gainer for the structure. [2]

WOOD ENERGY

Henry David Thoreau observed that wood heats twice; first when you cut it, second, when you burn it.

During the last half of the nineteenth century, cast iron

wood-burning stoves were the primary source of home heating. When less expensive, less tedious methods of heating became available, such as oil and coal burners, wood burning decreased. But today, with shortages and high prices of other home heating fuels, heating by wood is becoming a more attractive way to stay warm. It is estimated that one cord of dry wood is equal to a ton of coal or about 185 gallons of oil. Of course, accurate comparison assumes dry, properly seasoned wood and efficient heating units.

Comparing wood fuel with other energy sources, we find that using wood may be a favorable alternative. Wood produces less dangerous fumes than coal or gas. Ash, the residue from burning logs, is reusable as garden fertilizer. Wood is available in abundant quantities in the Northeast and is the only energy source that can be renewed. By cutting wood for fuel, we improve our woodlands as we thin them of diseased and undernourished trees.

When wood is properly seasoned, it performs best as an energy source. Protect wood from the weather to dry more efficiently by keeping it in some sort of shelter or shed, or even under a plastic tarp. When the logs are stacked, they should be arranged so that they have room to breathe. Another way to increase the rapidity of wood drying is to split the wood, exposing more surfaces to the air.

The design and capacity of the stove or furnace affect the amount of heat produced. Wood fires require direct attention to ensure efficient burning, and chimney fires may threaten if certain safety precautions are not taken. But, even as we consider the disadvantages of wood burning, we must acknowledge that wood is becoming a reasonable source of fuel. Because wood dealers are producing a high-quality product at economically competitive prices, wood is a good buy for home heating fuel. [3]

DRYING WOOD

Researchers at the Forestry Sciences Laboratory of the United States Department of Agriculture Forest Service, in West Virginia have come up with a few findings on drying fuel wood. In an effort to find the fastest and best way to dry wood, they cut ten cords of wood and banded them into two-foot by two-foot bundles. Each bundle was weighed at regular intervals to determine moisture content. Eight factors thought to affect drying time were tested: cutting season (spring or fall), species, length of stick, split versus unsplit, stacking versus ricking, piling in the woods versus piling in the open, and covered versus uncovered piles.

Perhaps the most interesting finding was that spring-cut wood, piled for six months, dried to an average moisture content of 22 percent. This was just 2 percent less than wood dried for twelve months under the same conditions. Splitting reduced moisture content by about 2 percent. Sticks that were twelve inches long lost about 1 percent more moisture than eighteen-inch sticks and 2 percent more than twenty-four-inch sticks. Ricking had no advantage over regular piling, but covering the piles did increase drying by several percentage points.

Red maple dried the quickest of the nine woods tested; going from 35 percent moisture content at cutting to fourteen percent after drying twelve months. White ash had the lowest moisture content at cutting time of the species tested, but lost the least during the drying period.

Drying was the slowest for unsplit wood, piled in the woods, and not covered. Wood actually took on moisture during some months under these conditions.

Most fresh cut hardwoods contain between 30 to 55 percent moisture content by weight. Reducing this moisture content to an ideal for burning — 17 percent — requires removal of up to fifteen hundred pounds of water per cord of wood. It is much better to let the sun and the wind do this drying than to make your stove do it.

This basic research project seems to say that wood cut by June first and properly handled will be quite burnable for the following winter, especially red maple.   [4]

WIND POWER

Did you ever stop to think that the same cool breeze that ruffles spring grasses and causes tulips and daisies to nod and bob might help to heat your house in the winter? In the 1920s and 1930s there were over six and a half million windmills on farms throughout the United States providing both water and electricity. Many now are abandoned and replaced by unsightly metal towers and tangles of wire. With the windmills went a chunk of the self-sufficiency that Americans are so proud of, which is one reason new homesteaders are turning once again to that free source of energy, the wind.

Homesteaders are not the only people reevaluating the possibility of wind power; scientists around the world are investigating new ways to harness that ancient Aeolian power. They seem to feel that the major potential for wind power lies not in small generators for houses and farms, but in large mills built where wind energy is maximal, such as on seashore cliffs or on tops of mountains.

The World Meteorological Organization calculates that twenty billion kilowatts of power are available around the world from wind, while others estimate up to eight hundred trillion kilowatts in the Northern Hemisphere alone.

It also stands to reason that the faster the wind is blowing, the more power we can derive from it. The power generated increases by the cube of the wind's speed. For example, we could get eight times as much wind power from a thirty-mile per hour wind as from a fifteen-mile per hour breeze. It is possible also to store energy in the form of electricity when the wind is not blowing. Combined with such a storage subsystem, windmills can provide us with self-contained sources for electricity on demand.

By far the most ambitious wind project is the Offshore Wind Power System proposed for the Gulf of Maine. This involves floating stations anchored by umbilical cords of electric cables that would carry the power generated by the windmills above the surface to the mainland. A storage system uses hydrogen gas prepared by electrolysis, which is stored in protected bladders or bags under the water. The hydrogen gas can be reconverted to electricity when wind production falls below demand. Such a system is no more technically complicated than existing offshore oil rigs, but it is far less polluting and does not deplete any natural resource. Theoretically, similar stations along the east and west coasts could provide all the electrical needs of the nation. In our search for new sources of energy and ways to discontinue polluting and despoiling our natural resources, we might do well to pay attention to which way the wind is blowing. [5]

WIND-POWERED GENERATING SYSTEMS

With the cost of electricity rising, you may want to consider supplementing your home or business power supply with a wind-powered generating system. Or, if you live on an island or own a remote camp, you may want to generate all your power from the wind.

A wind plant, or wind turbine or wind generator, consists of a tower on which is mounted a rotor and a tail vane, a Direct Current generator or Alternating Current alternator, and a storage system, which is usually a battery. The storage system is necessary because energy output depends on wind velocity, which is variable. Few wind plants can provide all the electrical demands of a home or business, but as supplemental sources of power, wind plants are becoming more and more cost effective.

The effectiveness of a wind plant depends on its site and on the height of the tower. The proper height is thirty feet higher than the nearest obstacles within a three hundred-foot radius in all directions. A ninety-foot height is recommended wherever possible.

The tower itself can be a tapering, self-supporting design, or a less expensive and nontapering tower held by guy wires sunk into concrete. In most cases your site will determine your choice of tower.

Determining whether your site is appropriate is the first step in choosing a wind plant. Both wind speed and turbulence are crucial factors. A wind tower must be located so that it is in a relatively turbulence-free airstream. Planetary, or prevailing, winds show constant directional characteristics, such as those on an exposed hill or shoreline. Local winds, in contrast, are caused by temperature variations and local conditions. The best site is one where local wind patterns reinforce planetary patterns.

Ground level wind speed is not a good indicator, for friction with the earth's surface causes drag. Speed increases proportional to the distance from the ground, so a general rule of thumb is that the higher your tower, the higher the wind speed.

To evaluate your site, try streaming a long ribbon from a guyed sapling or tall pole. For two weeks, observe whether it streams evenly in high winds from varying directions. You can measure wind velocity with a simple and inexpensive wind meter and should chart a daily average for two weeks. If ground speed is less than eight miles per hour, and commercial power is readily available, a wind plant is not worth the investment. Also, try checking with your local weather station for help in forecasting wind distribution. [6]

KEROSENE

Kerosene is the standard fuel for wick lamps. The term is used loosely to describe a thin, flammable oil with a rather high ignition or flash point, roughly 160 degrees F. That high ignition temperature makes kerosene safer for household use than more volatile alcohol or gasoline—in fact, a lighted match can be dunked in a pot of kerosene without igniting it.

Kerosene can be produced from coal oil, oil shale, and wood fiber, but most commonly it is refined from petroleum. It can also be made in different grades. The key to understanding lamp oil is that not all kerosene is created equal.

What you need for clean lamplighting is a special low-sulfur kerosene usually referred to as Number 1-K. It is also sometimes called Type A, paraffin oil, or water-clear or lighting-grade kerosene. Number 1-K is the champagne of kerosene, without the bubbles. Number 2-K kerosene, which is a lower grade and much less expensive, can also appear clear, but it contains about ten times as much sulfur as 1-K. Be careful not to purchase other fuels that may be passed off as kerosene—fuels such as Number 1 fuel oil, Number 1 diesel fuel, and jet engine fuel.

The trouble with using fuels other than 1-K is that they contain too much sulfur and tar. Gradually, the tar gums up the wick, reducing capillary action. The more low-grade fuels you burn, the dimmer the light and the smokier the flame. If you're compelled to burn 2-K, for example, it's necessary to clean the wick frequently—or replace it—and also to clean the reservoir. That still won't increase the amount of light, however, or reduce the unpleasant odor when the lamp is burning. A wick lamp is not fuel-efficient; it burns only about 70 percent of the fuel, and if the unburned kerosene vapor is loaded with sulfur, you smell it. Switching to 1-K won't boost efficiency much, but it sure cleans up emissions. [7]

EFFICIENT LIGHTING AND SMALL APPLIANCES

Through the wonder of electricity, we literally can turn night into day. The electric light, used for home lighting and home decorating, safety, and security, is an essential part of our lives, yet extravagant use far outstrips basic needs.

Lighting amounts to 3.5 percent of a home's total energy bill and 8 to 16 percent of the total electric bill. By using and maintaining efficient lighting and by turning off lights when they are not needed, it is possible to save energy and dollars.

Light bulbs are rated in lumens (the light output) and watts (the rate at which they consume electricity). Today, most packaging also indicates the average life of bulbs, usually 750 to 1,000 hours.

Long-life bulbs, which last an average of 2,500 hours, give off 20 percent less light per watt than standard bulbs. Long-life bulbs are not energy-efficient; use them only in areas where it is difficult to replace them, such as in stairwells.

If higher illumination is required, consider using one high-wattage incandescent bulb instead of several bulbs of low wattage. For example, one hundred-watt bulb produces approximately the same amount of light as two sixty-watt bulbs, yet uses 20 percent less energy. For safety, never leave live light sockets empty. Fill them with burned-out bulbs. Never use a higher-wattage bulb than specified for a fixture. Increased wattage may cause the fixture to overheat—a fire hazard—and shorten bulb life.

Fluorescent lights are three to five times more efficient than incandescent bulbs and last seven to ten times longer. They also generate less heat, a plus during summer months. One forty-watt fluorescent tube provides 3,200 lumens, whereas one hundred-watt standard incandescent bulb provides 1,750 lumens. Use fluorescent lighting in the kitchen, bathroom, laundry, workshop, and recreation or family areas of your home. [8]

COAL

Coal commenced to be formed soon after the first plants established themselves on land in the Devonian period, almost half a billion years ago. However, the great coal-making time commenced in the Carboniferous period 225 to 350 million years ago. The climatic, geologic, and biologic conditions that encouraged its creation might be described by attempting to recapitulate a particular place and moment of Carboniferous time.

Take, for example, one which has become almost synonymous with coal: the Pennsylvanian period or the Upper Carboniferous period in, say, the year 300,000,000 B.C. and in the area of northeastern Pennsylvania near the present Wilkes-Barre. The climate was essentially subtropical—muggy, rainy, frostless, and virtually seasonless. The terrain did not remotely resemble the gnarled mountain country which exists there now. It was very flat, only a few feet above the level of a vast, shallow sea which to the westward covered much of what is now the central United States. The land itself was composed of layers of sandstones and clays, either dampish or covered by thin sheets of water that was fresh but stagnant. The few rivers that flowed did so sluggishly.

This warm, wet place and time produced vegetative growth as perhaps no others have in the history of Earth. The land was covered by immense and dense jungles. Impressive as they were in terms of sheer quantity of organic matter, they probably would be regarded by present botanical and esthetic standards as fairly monotonous. Evolutionary forces had not had long to work on terrestrial forms, and the flora was composed of a relatively few species of simple but inconceivably numerous plants. Among the most abundant were the calamites, which quite closely resembled their surviving descendants, the humble horsetails, except that calamites grew as trees, their slender jointed trunks towering over the swamps. Perhaps the largest trees of the jungles were cordaites, forerunners of our contemporary conifers. They stood more than a hundred feet tall and bore straplike, three-foot-long leaves. Mile after mile was covered with true ferns, which grew in the manner of shrubs, vines, and also trees. Along with the ferns, a low bush with vaguely laurel-like leaves, constituted most of the understory. There were no flowering plants.

A critical characteristic of the Carboniferous period was its environmental stability, the hothouse conditions prevailing for 125 million years. There were no glaciers, no exceptional droughts, no major terrestrial upheavals or eruptions. For millions upon millions of years the jungles continued to absorb solar energy, the individual plants living, growing, and dying. Of the enormous tonnage of leaves, twigs, trunks, and other vegetative debris that fell into the swamps, ninety-eight percent decomposed quickly into its component elements. The two percent that did not and was preserved provided the world with trillions of tons of coal. [9]

OIL-PLATFORM FIRES

Life aboard an offshore oil-well platform is hardly serene. Storms, accidents, and fires are ever-present dangers. While wave action still presents the greater risk to the people who toil at the drilling rigs, nothing is more feared at an oil-well site than a fire. To protect workers from this dreaded peril, seminars are held, drills are performed, life-support devices are provided, and immense expenses are incurred in bringing fire-fighting vehicles to the scene of action. Yet fires continue to exact a deadly toll.

From only 1,000 operating off-shore drilling rigs in 1971, the number soared to 4,000 in 1981. The Hughes Tool Company estimates that this year [1983] the number of rigs in use will climb to 4,500, and drilling expenditures will increase to $41 billion. From Alaska's North Slope to Indonesia and Java, offshore drilling is a growing operation. It is an industry with great potential—and grave dangers from accidents, especially off the continental shelf worldwide. Prevention of accidents in this industry has occupied many minds, including those of the conservation division of the U.S. Geological Survey. Every eighteen months it conducts a seminar, and its published technical reports disseminate ideas for the control of blowouts and fires on offshore platforms. Prevention and suppression are the keynotes.

A blowout occurs when the pressure exerted by a column of drilling mud (used to lubricate the bit) is less than that of the oil or gas formation penetrated by the well. When fluid from that formation enters the well and forces the drilling mud up the casing, a blowout results. If gas or oil escapes, friction, chemical reactions, or other agents can cause the mixture to ignite. To stop a threatening blowout, caps called blowout preventers must be closed. Then, to remove the oil or gas from the well column, the density of the drilling mud must be increased by an emergency high-pressure flow line and adjustable choke. When the well is located offshore, the blowout preventers may be positioned on the sea floor rather than at the surface, meaning that operators must reach underwater in person or by mechanical means to control the threatening blowout.

Sometimes such disasters defy control. Fires resulting from blowouts can destroy a platform or rig in short order, for temperatures may rise as much as 2,000 degrees F. [10]

ARCTIC MINING

It seems as if no physical problems encountered in the Arctic environment are insurmountable. A new generation of technology, engineering, and ship design has emerged in the pursuit of Arctic riches. Assuming vast oil deposits are found and determined to be retrievable on favorable commercial terms, we are on the verge of a mammoth effort to exploit Alaskan resources. Never before has the depth of interest in operating in the U.S. Arctic been so great. Nor have the sums committed been so large—they are already nearly $20 billion, with far more to come. Exxon USA estimates more than $300 billion may be spent on Alaskan petroleum resources. The Canadian Arctic prospects, which may be as large, have involved commitments in the range of $5 billion to $10 billion, with more anticipated.

Thus the prospects and the investments in the U.S. Arctic have their parallel eastward, with very large Canadian undertakings in Canada's portion of the Beaufort Sea, and further eastward to the Arctic Islands, where the existence of substantial deposits of natural gas is already established. Moreover, in the U.S. Arctic, immense riches of coal (perhaps 130 billion tons) and other hard minerals (gold, silver, iron ore) reinforce the long-held conviction that the resources of these lands—including timber—are great untapped assets; the Canadian Arctic is thought to be one of the largest and richest of geological regions.

In both the United States and the Canadian Arctic zones, the greatest gamble today is probably not in the actual hunt for recoverable resources, nor with what most people think is a fragile environment, but in the troubled economies and reduced energy needs of both countries. It is entirely possible that Arctic developers will never quite realize their great expectations for lack of a large enough commercial incentive.

Nevertheless, enough has happened already to bring

the historic dream of exploiting the Arctic far beyond its living resources closer to reality; of the goal of making passages east and west through ice and heavy seas; of finding ways to work year-round; of the fashioning of special means to penetrate the frozen earth to great depths and then to recover the resource; of learning much more about the Arctic environment, its weather, its capability to recover from pollution. [11]

The Nuclear Sword

REDUCTION OF THE OZONE LAYER

Although pure ozone is toxic to both plants and animals, its presence high in the atmosphere is essential to absorb biologically harmful ultraviolet radiation from the sun. In fact, the evolution of a protective ozone shield was a necessary condition for the appearance of terrestrial animals and plants.

Recent studies have shown that the huge quantities of nitric oxide that would be injected into the atmosphere during a nuclear war could reduce the ozone shield's effectiveness against ultraviolet radiation in the higher stratosphere by as much as 50 to 70 percent, increasing ozone levels in the lower atmosphere, where it is toxic, to more than 160 parts per billion (about five times the normal level). The ozone increases in the lower atmosphere (troposphere) would kill plants and cause respiratory distress and disease in animals, while only partly offsetting the large increases in solar ultraviolet radiation pasing through the reduced ozone shield. In the worst of the nuclear war scenarios, levels of biologically effective ultraviolet radiation reaching the earth's surface would increase by 500 to 1,000 percent.

Ultraviolet radiation is absorbed by the proteins and

nucleic acids of living cells, resulting in a variety of types of cell damage, including mutation. The phytoplankton living in the top few meters of ocean water have been found to be particularly sensitive, as are many higher orders of plants. Fish are also sensitive. Scientists estimate that a 16 percent reduction in the ozone layer would kill up to half the anchovies in the top ten meters of the clearest ocean water, or else force them to go deeper.

Long-term climatic changes also are a possible consequence of a major nuclear war. The deposition of soot from the many fires in cities, forests, and oil and gas fields could cause polar snow and ice to melt.

Changes in the temperature structure of the atmosphere caused by the injection of soot and dust and the reduction of the ozone layer could also affect climate. Geologists believe that for more than 99 percent of the history of the earth there have been no polar ice caps. Our present climatic situation could be considered abnormal, in which case an appropriate stimulus might cause a reversion to the more normal situation—the melting of the polar ice. Alternately, the earth could be plunged into a deeper Ice Age, such as existed about twenty thousand years ago. Unfortunately, not enough is known about climatic change to make accurate predictions. [1]

A DARKENED WORLD

The detonation of tens of thousands of nuclear weapons would ignite tens of thousands of mass fires in cities, industrial centers, gas and oil fields, fossil-fuel stockpiles, refineries, military installations, croplands, and forests. Fires and firestorms caused a major portion of the damage to the cities of Hiroshima and Nagasaki. Detonation of nuclear weapons over gas and oil fields would result in numerous blowouts, and without expert personnel to extinguish the resulting fires these wells would burn as long as positive pressure remained.

Photosynthesis in plants would cease, because sunlight would be blocked by the black smoke carried to high altitudes and by the hundreds of millions of tons of fine dust thrown into the atmosphere by surface bursts. Most of the phytoplankton and herbivorous zooplankton, the microscopic organisms that begin the food chain for all marine animal life, would die in the oceans of the Northern Hemisphere. Many scientists now agree that it is likely the noontime sky would be as dark as a moonlit night for up to three or four months following a nuclear exchange. The surface of the earth would cool, as most of the incoming solar heat would be absorbed in the atmosphere, and the change in the temperature structure of the atmosphere would result in unpredictable changes in global weather.

A similar scenario is now the accepted explanation for the massive extinction of the dinosaurs and many other plant and animal species sixty-five million years ago. The argument, based on geochemical studies, is that the collision of a ten-kilometer-diameter asteroid with the earth resulted in the injection of large quantities of fine dust into the atmosphere, causing the darkening and cooling of the planet.

Animals and plants surviving the vast initial destruction of a nuclear war caused by the blast, heat, and nuclear radiation would be subjected to three or four months of the nearly total absence of sunlight, with resulting freezing temperatures and extreme weather conditions. As the fires ceased to burn and the soot settled from the atmosphere, extremely high levels of solar ultraviolet radiation would begin to penetrate to the biosphere, doing additional damage to plant life and blinding terrestrial animals. [2]

A FULL-SCALE NUCLEAR WAR?

In an all-out nuclear war, thousands of bombs would be exploded on civilian and military targets all over the Northern Hemisphere. It is almost impossible to comprehend what this would mean for humans. Comparisons to past disasters are totally inadequate.

We must realize that no outside help would be available. Systems of transportation, communication, and power would be disrupted totally. Food, water, and electricity would be unavailable. Televisions, radios, and telephones would be useless.

Survivors would have little or no medical care to ease their suffering. Hospitals, antibiotics, blood banks, and narcotics for pain would be destroyed—to say nothing of doctors and nurses, themselves killed or crippled.

Some creatures, however, would thrive. Many insects and disease-carrying pests that are resistant to radiation would flourish in the aftermath of nuclear explosions. Rats would multiply. Plagues would fester and spread, abetted by huge numbers of untended human and animal corpses.

Finally, we must consider the possibility of irreversible damage to nature's fragile systems. For example, the ozone layer in the atmosphere might be so seriously depleted that many animals and plants would die, unprotected from the sun's powerful ultraviolet rays. Birds, bees, and other vital links in the food chain could perish, causing the ecosystem to collapse—perhaps forever. If this happened, it is difficult to imagine how human life could continue.

As weapons continue to consume an ever larger share of the planet's resources, poor people everywhere suffer the most. Some, in frustration, turn to violence. Thus, the arms race not only threatens future violence, but also it guarantees violence now.

Parents and Teachers for Social Responsibility in their publication "What About The Children?" tell us that many children worry constantly about their prospects of growing up. They tend to believe that nuclear weapons are here to stay and that war is inevitable. For these children, the future is hopeless, and adults seem powerless to change the situation. In fact, children often wonder whether the adults are even trying.

We are told that nuclear weapons protect our freedoms and beliefs. Instead, we see these weapons eroding the very ideals they are supposed to defend. They undermine our sense of human dignity. They steal hope from our children.

Nuclear weapons represent a new and terrible form of tyranny. Daily we live as hostages to the stark reality that a single person, or even a computer accident, can destroy everything we know and love. [3]

WHAT ABOUT THE CHILDREN?

I have talked about the results of a nuclear war on our wildlife and our environment. But I have grandchildren; what about them? Nuclear weapons have been used twice on human populations—on Hiroshima, August 6, 1945, and three days later on Nagasaki. While today's weapons are fifty to twelve hundred times more powerful, it is instructive to review the effects of those first bombings, especially what happened to the children.

The immediate effect, of course, was the instant incineration of many thousands of human beings. Thousands more died within days or weeks. And then there were the survivors. Experts have studied both the short- and long-term impact of the Hiroshima–Nagasaki bombings. They tell us that the survivors who suffered most were babies, children, the aged, the sick, and the handicapped—in short, those most in need of strong family support.

At the same time, families disintegrated. Key members died. Others were lost. Some, out of hopelessness, abandoned one another and wandered about as if in a dark dream. It is reported that during the days following the bombings some of the orphans placed signs in the rubble where their homes had stood. "Mother, where are you?" "Father, where are you?"

Radiation most seriously damages cells that are actively growing and multiplying. It is particularly damaging to infants still in the womb. In Hiroshima and Nagasaki, many pregnant women suffered spontaneous miscarriages. Others delivered babies marked by a variety of physical and mental defects, many born with small heads and underdeveloped brains. Other abnormalities included cirrhosis of the liver, funnel chest, and mongolism, to name but a few.

Children also contracted leukemia and other cancers. Strontium 90, often called a bone-seeker, was a by-product of the nuclear explosions. It contaminated milk and then made its way into the bones of children, where it was stored like calcium. Months or years later, children developed leukemia.

Physical diseases and deformities were accompanied by a variety of psychological scars and illnesses. These effects continued for many years in the form of chronic depression, diminished vitality, and high levels of anxiety.

The list of problems goes on and on. It has been said that human beings are the soft targets in a nuclear war. If this is so, then children are the softest targets of all.

The Hiroshima and Nagasaki bombs were tiny compared to the bombs of today. Also, the survivors of those bombs received assistance from the outside world. There would be no such help in the next nuclear war. There would be no outside world. The children instantly vaporized probably would be the lucky ones. [4]

TWISTED WORDS, DISTORTED IMAGES

Language plays a subtle but important role in maintaining the arms race. Nuclear weapons are given appealing names, such as Little Boy, Peacekeeper, Badger and Honest John.

Governments speak of "nuclear shields" and "nuclear umbrellas," as if there were plans to stop incoming missiles. In fact, no such plans exist.

One government publication, written for children, describes nuclear war as if it were just another social inconvenience. The text then assures us: "Everyone will work together to help the community get back to normal."

The term "terrorism" is used when a bomb is sent through the mail or hostages are held for ransom. Yet when governments threaten to kill millions of people with nuclear weapons, that is not called terrorism. . . . It is called strategy.

Even the term "arms race" is misleading. It implies that the side with the most weapons will be a winner.

When language is distorted in this way, our perceptions of reality are also distorted. For example, the term "acceptable casualties" leads us to think of a small number, when in fact the term is used to describe the incineration of millions of human beings and perhaps an entire civilization. A so-called limited war in Europe would cause more terrible destruction than any war in human history.

At the head of this parade of misleading images, we find "defense," "security," and "national interest," where there is no defense, diminishing security, and a very serious question about whose interests, if any, are being served.

We must resist such false and twisted thinking and use words, phrases and images that convey the reality of mass destruction. . . . [5]

THE ULTIMATE FOLLY

Anyone concerned about the environment cannot remain aloof from the growing debate over nuclear weapons. Today the world has a stockpile of such weapons with a total destructive power one million times greater than the bomb that destroyed Hiroshima.

"How would the human species make out in a nuclear war?" someone asked an expert on this issue at an Audubon convention. The answer: "Human beings would do worst. Cockroaches would do best."

It's estimated that a nuclear war would cause at least a billion immediate deaths, with another billion injuries from blast, heat, and radiation. This casualty list includes a good share of the earth's human population, primarily residents of the Northern Hemisphere. But what about those not killed right away? What do they have to look forward to?

Carl Sagan and Paul Ehrlich note that the two most significant effects of nuclear war between this country and the Soviet Union would be widespread darkness and cold. Vast firestorms would send millions of tons of ash and soot aloft. Sunlight would be reduced by 95 percent or more; temperatures would drop at least thirty-five degrees; photosynthesis would slow or stop; virtually all land plants in the Northern Hemisphere would be damaged or killed. In this frigid, twilight world, starvation and disease would afflict many of the surviving creatures. Only the scavengers that could withstand extreme cold would be likely to flourish, thanks to the millions of unburied human and animal bodies. And when the dust settled, the ultraviolet radiation from the sun, penetrating the depleted ozone layer, would kill many of those species still extant.

"A realistic look at the earth after a nuclear attack leaves us guessing that a quick, merciful roasting in a personal fireball might be a better way," observed George Woodwell, a noted marine biologist.

In Boulder, Colorado, Professor Gilbert White and his students have just finished a postwar scenario for Boulder County depicting the effects on local life-support systems of a nuclear detonation in nearby Denver. This report should make local opinion leaders and news media take notice, for it brings the issue home in dramatic detail, pinpointing the impact of a nuclear exchange on the country's air, water, and land, including the farmland on which the local food supply would depend.

In every way possible, we must work for the eventual elimination of all nuclear weapons. Early conservationists once saved the plumed birds of this country from extinction. Now they have the opportunity — and the obligation — to help save all life in all countries from extinction. [6]

Contributors

THE SEASONS
1. Seasons. Jane P. Curtis.
2. A Hillside in Spring. Nancy Martin, Vermont Institute of Natural Science.
3. Spring Trees. Nancy Martin, Vermont Institute of Natural Science.
4. Rite of Spring. John H. Mitchell, Massachusetts Audubon Society vol. 18 #9, May/June 1979.
5. Spring Sounds. Jenepher Lingelbach, Vermont Institute of Natural Science.
6. Shadbush. Jane P. Curtis.
7. Lilac Time. *Rutland Vermont Daily Herald*, May 20, 1974.
8. Leaving Wildlife Youngsters Wild. Vermont Fish & Wildlife Department.
9. Dandelions. Jenepher Lingelbach, Vermont Institute of Natural Science.
10. Dew. Jenepher Lingelbach, Vermont Institute of Natural Science.
11. Hiking. Jane P. Curtis.
12. Stars on a Summer Night. Jane P. Curtis.
13. The Scythe. Will and Jane Curtis.
14. Bright Passage. Bill Vogt. Copyright 1977 by the National Wildlife Federation. Reprinted from Oct./Nov. issue of *National Wildlife Magazine*.
15. The Maple Casebearer. Jenepher Lingelbach, Vermont Institute of Natural Science.
16. Ponds Change in Autumn. Michael Caduto, Vermont Insitute of Natural Science.
17. Banking the House. Barney Crosier, *Rutland Vermont Daily Herald*.
18. Winter Bird Feeding. Mary Richards, Vermont Institute of Natural Science.
19. Snow Is Nature's Security Blanket. Ned Smith. Copyright 1973 by the National Wildlife Federation. Reprinted from Feb./March issue of *National Wildlife Magazine*.

20. Snowshoes. Will and Jane Curtis.
21. Animals Keeping Warm. Nancy Martin, Vermont Institute of Natural Science.
22. Winter Feet. Jenepher Lingelbach, Vermont Institute of Natural Science.
23. Snow Fleas. Anette Gosnell, Vermont Institute of Natural Science.
24. A Long Winter's Nap. Margaret Barker, Vermont Institute of Natural Science.
25. "Habitat" Gardening. Nancy Withington, Vermont Institute of Natural Science.

PLANTS & GARDENS

1. How Soils Are Made. Extension Service, University of Vermont.
2. Roadside Flowers. Jenepher Lingelbach, Vermont Institute of Natural Science.
3. Weeds: Foreign Friends or Alien Enemies? Jenepher Lingelbach, Vermont Institute of Natural Science.
4. Poisonous Plants. *Montshire Museum of Science Newsletter*, August 1977.
5. Creeping Menace. *National Wildlife Federation News Release*.
6. Ancestral Diet. Jane P. Curtis.
7. Garibaldi's Garden. Jane P. Curtis.
8. High-yield Vegetables. "Questions and Answers." Elsie Cox. March 1982. Reprinted courtesy of *Horticulture: The Magazine of American Gardening*, 755 Boylston St., Boston, MA 02116. Copyright © 1982, Horticulture Associates.
9. How to Make Better Compost. Extension Service, University of Vermont.
10. Manure Means Money. Extension Service, University of Vermont.
11. Witch Hazel. Mary Richards, Vermont Institute of Natural Science.
12. Garden Wrap-up. Extension Service, University of Vermont.
13. Home Storage of Vegetables. Extension Service, University of Vermont.
14. Leaf Drop of Evergreens. Extension Service, University of Vermont.
15. Autumn Leaves. Jane and Will Curtis.
16. Fall Lawn Care. Extension Service, University of Vermont.
17. Watering and Indoor Plants. Extension Service, University of Vermont.
18. Moss and Berry Bowls. Extension Service, University of Vermont.

MAMMALS

1. Animals. Douglas Newton, *Metropolitan Museum of Art Bulletin*, Fall 1981. Copyright © 1981 The Metropolitan Museum of Art.
2. Growing Old. *National Wildlife Federation News Release*.
3. Yawning Animals. *National Wildlife Federation News Release*.
4. Animal Language. "Out of the Mouths of Beasts." Carol Grant Gould, April issue *Science '83*. Reprinted by permission of *SCIENCE 84*

Magazine, © the American Association for the Advancement of Sciences.

5. Insulation. Jenepher Lingelbach, Vermont Institute of Natural Science.
6. Migration. "Migration: The Irresistible Summons." Cathy Jarman. Copyright 1973 by the National Wildlife Federation. Reprinted from Sept./Oct. issue of the *International Wildlife Magazine*.
7. The Hunter and the Hunted. Judith Irving, Vermont Institute of Natural Science.
8. Bag Balm. "More Than Just an Udder Solution." Steve Fustero. July/Aug 1983. Reprinted by permission of *SCIENCE 84 Magazine*, © the American Association for the Advancement of Sciences.
9. The Beaver. Blair C. White, *Conservation Digest*, South Dakota Dept. of Game, Fish & Parks.
10. Red Squirrel. Richard Headstrom, Vermont Institute of Natural Science.
11. Antlers. "Wildlife Sketchbook." Ned Smith. Copyright 1966 by the National Wildlife Federation. Reprinted from Aug./Sept. issue of *National Wildlife Magazine*.
12. The Red Deer. *National Wildlife Federation News Release*.
13. Himalayan Animals. *National Wildlife Federation News Release*.
14. The Buffalo Graze, Again. "Saved Just in Time: The Buffalo Graze Again on Our Plains." John G. Mitchell, *Smithsonian Magazine*, May 1981.
15. Mules. Peter Chew, *Smithsonian Magazine*, November 1983.
16. Meet Wildlife Enemy No. 2. Roger Caras. Copyright 1973 by the National Wildlife Federation. Reprinted from Feb./March issue of *National Wildlife Magazine*.
17. Coyotes. Wendy Vogt, Vermont Institute of Natural Science.
18. The Big, Good Wolf. Scott Barry. Reprinted with permission from *DEFENDERS Magazine*, February 1981.
19. A Flash of Mink. Wayne Handley, *Massachusetts Audubon*, April 1979.
20. The Weasel. Mary Richards, Vermont Institute of Natural Science.
21. The Fisher. Donald Wharton, *American Forests*, September 1983.
22. The Water Shrew. "Little Devil." Jeri Gailbraith. Copyright 1983 by the National Wildlife Federation. Reprinted from Aug./Sept. issue of *National Wildlife Magazine*.
23. The Jaguarundi. Roxanna Sayre. May 1983. Reprinted from *AUDUBON*, the magazine of the National Audubon Society; Copyright © 1983.
24. The Cougar's New Cloak. Gary Turbank. Copyright 1982 by the National Wildlife Federation. Reprinted from April/May issue of *National Wildlife Magazine*.
25. Nature's Golden Racer: The Cheetah. Emily and Ola d'Aulaire. Copyright 1971 by the National Wildlife Federation. Reprinted from Jan./Feb. issue of *International Wildlife Magazine*.

26. The Wiley, Indigestible Armadillo. *National Wildlife Federation News Release.*
27. The Sea Otter. Vermont Institute of Natural Science.
28. Animals' Diving Gear. *National Wildlife Federation News Release.*
29. The Tragic Facts About Whales. Courtesy of the Whale Protection Fund Center for Environmental Education.
30. The Bowhead Whale. Courtesy of the Whale Protection Fund Center for Environmental Education.
31. White Whale. "The White Whale." Wayne Hanley, *Massachusetts Audubon Society*, August 1980.
32. Questions for a Blue Whale. Russ Kinne. March 1980. Reprinted from *AUDUBON*, the magazine of the National Audubon Society; Copyright © 1980.

BIRDS

1. Rites of Spring. Mary Richards. Vermont Institute of Natural Science.
2. Growing Up in a Nest. Jane and Will Curtis. *Welcome the Birds to Your Home.* Stephen Greene Press, 1980.
3. Masters of Adaptation. Ibid.
4. Spring Bird Migration. *Sanctuary*, Massachusetts Audubon Society, April 1981.
5. Feathers. David Cavagnaro. *Feathers.* Illustrated by David Cavagnaro and Frans Lanting, Graphic Arts, Portland, Oregon, 1982.
6. How Birds Change Their Clothes. Betty Backus, *Wild Bird Guide* Summer 1982, Bird Friends Society.
7. Man-made Bird Houses. Jane and Will Curtis. *Welcome the Birds to Your Home.* Stephen Greene Press, 1980.
8. "Mobbing." "How Birds Use 'Mobbing' to Subdue Their Enemies." John V. Denis. *Wild Bird Guide* Fall 1981, Bird Friends Society.
9. Independent Nestlings. Ned Smith. Copyright 1973 by the National Wildlife Federation. Reprinted from June/July issue of *National Wildlife Magazine.*
10. The Upside-Down Bird. "The Upside Down Bird." Julie Remick. *Wild Bird Guide* Summer 1983, Bird Friends Society.
11. Whip-Poor-Wills. Fran Howe, Vermont Institute of Natural Science.
12. Bluebirds. Vermont Institute of Natural Science.
13. Cukoos Aren't Crazy, Just Lazy. *National Wildlife Federation News Release.*
14. Swifts. Nancy Martin, Vermont Institute of Natural Science.
15. The Pine Siskin. V. L. Miller, *Wild Bird Guide* Winter 1983, Bird Friends Society.
16. Grouse. Nancy Withington, Vermont Institute of Natural Science.
17. Raising a Great Horned Owl. Michael Caduto, Vermont Institute of Natural Science.
18. Ostriches. "Speedy Ostriches Outrun Lions, But Can't Escape Civilization." *National Wildlife Federation News Release.*

19. The Bald Eagle. Adapted from *AUDUBON*, January 1984 issue.
20. Condors. "Bugged Condors Reveal Secrets." *Action Alert*, National Audubon Society, February 1983.
21. Turkey Vultures. *Nature's Way*, Massachusetts Audubon Society, Summer 1980.
22. Kingfishers. Jenepher Lingelbach, Vermont Institute of Natural Science.
23. The Common Loon. Peggy Milardo, *Maine Audubon Society Quarterly*, Summer 1983.
24. The Bittern. Hank Fischer. Reprinted with permission from *DEFENDERS* magazine. Oct. 1981.
25. Whooping Cranes. Jane P. Curtis.
26. Gulls. "Gull as in Gullet." Jack Denton Scott. Copyright 1974 by the National Wildlife Federation. Reprinted from Feb./March issue of *National Wildlife Magazine*.
27. Puffins. Leni Sinclair, *National Wildlife Federation News Release*.
28. Killdeer. Mary Richards, Vermont Institute of Natural Science.
29. A Guide for Gannet Gazers. George Harrison. Copyright 1981 by the National Wildlife Federation. Reprinted from May/June issue of *International Wildlife Magazine*.

INSECTS

1. Insect Structure. Jane P. Curtis.
2. Insects in a Stream. Michael Caduto, Vermont Institute of Natural Science.
3. Insect Antifreeze. "The Bugs of Winter." Joe Allen Tuthill. Jan./Feb. 1983. Reprinted by permission of SCIENCE 84 Magazine, © the American Association for the Advancement of Sciences.
4. A Bounty of Beetles. Bernard Durwin and Gerhard Scherer. Nov. 1981. Reprinted from *AUDUBON*, the magazine of the National Audubon Society; Copyright © 1981.
5. Beetlemania. Douglas J. Preston. With permission from *Natural History*, Vol. 92, No. 5; Copyright the American Museum of Natural History, 1983.
6. Honeybees. "Honey Bees." Sami Izzo, Vermont Institute of Natural Science.
7. Bumblebee. Jane P. Curtis.
8. Bees As Undertakers. "A New Undertaking For Honeybees." November 1981. Reprinted by permission of SCIENCE 84 Magazine, © the American Association for the Advancement of Sciences.
9. Bee Bites. Roger B. Swain. June 1980. Reprinted courtesy of *Horticulture: The Magazine of American Gardening*, 755 Boylston St., Boston, MA 02116. Copyright © 1980, Horticulture Associates.
10. Lightning Bugs. Jenepher Lingelbach, Vermont Institute of Natural Science.
11. Grasshoppers. Ibid.
12. Crickets. Mary Richards, Vermont Institute of Natural Science.

13. Butterflies of Spring. Nancy Martin, Vermont Institute of Natural Science.
14. The Monarch. Wayne Hanley, Massachusetts Audubon Society, August 1978.
15. Pill Bug. Jenepher Lingelbach, Vermont Institute of Natural Science.
16. Aphids. Mary Richards, Vermont Institute of Natural Science.
17. Crab Spiders. Margaret Barker, Vermont Institute of Natural Science.
18. Black Widow Spider's Bite. *National Wildlife Federation News Release.*

WATER & AQUATIC LIFE
1. Water. Massachusetts Audubon Society, February 1979.
2. Water Tables. Educational Services, National Audubon Society.
3. Woodlots and Water. Richard Brett. *The Country Journal Woodlot Primer.* Reprinted by permission from *Blair & Ketchum's Country Journal,* Copyright © 1983, Historical Times, Inc.
4. Trends in Neurobiology. William J. Adelman, Jr. *Oceanus,* Summer 1983, Woods Hole Oceanographic Institute.
5. Leaking Gas Tanks. Marcel Moreau, *Habitat,* January 1980, Maine Audubon Society.
6. Wetlands. Public Service Information. Massachusetts Audubon Society.
7. Swamp Bubbles. Michael Caduto, Vermont Institute of Natural Science.
8. Peat. "Peat: Ireland's Cheap Heat." Kit and George H. Harrison. Copyright 1982 by the National Wildlife Federation. Reprinted from Jan./Feb. issue of *National Wildlife Magazine.*
9. Tidepools. "Poisonous Tide Pools." *Oceanus,* Summer 1982, Woods Hole Oceanographic Institute.
10. Estuary. From "Estuary," Copyright 1972 by the National Wildlife Federation, 1412 16th Street, NW, Washington, DC 20036. Revised 1983.
11. Salamanders. "The Woods Are Alive with Salamanders." Harry Ellis. Copyright 1977 by the National Wildlife Federation. Reprinted from Oct./Nov. issue of *National Wildlife Magazine.*
12. Shells and Man. *Montshire Museum of Science Newsletter,* Summer 1980.
13. Starfish. *National Wildlife Federation News Release.*
14. Sea Urchins. "Today's Diner, Tomorrow's Dinner." George S. Fichter. Copyright 1979 by the National Wildlife Federation. Reprinted from the March/April issue of *International Wildlife Magazine.*
15. Blue Lobsters? *National Wildlife Federation News Release.*
16. Pickled Wrinkles. *Maine Audubon Society Quarterly,* Fall 1982.
17. The Living Fossil. "The Living Fossils Habits." Dan and Vicki Gibson, *Cape Naturalist,* Summer 1983, Vol. 12 No. 1. Reprinted by permission of The Cape Cod Museum of Natural History, Inc.

18. Sand Dunes. "Sand Dunes are Important to Maine's Beaches." Marcel Moreau. Maine Audubon Society.
19. The Science of Oceanography. Charles D. Hollister. *Oceanus*, Summer 1983, Woods Hole Oceanographic Institute.
20. Salmon. Ted Willims, *Smithsonian*, November 1981.
21. Return to the Connecticut. Will Curtis.
22. The Seahorse. "From Father Neptune's Stables." Lowell P. Thomas. Reprinted by permission from *Sea Frontiers* © 1982 by the International Oceanographic Foundation, 3979 Rickenbacker Causeway, Virginia Key, Miami, Florida 33149.
23. Shrimp Farming. "Intensive Shrimp Culture in Japan." Daniel Spotts. Reprinted by permission from *Sea Frontiers* © 1983 by the International Oceanographic Foundation, 3979 Rickenbacker Causeway, Virginia Key, Miami, Florida 33149.

PLACES

1. Canyon de Chelly. Jane P. Curtis.
2. Early Prairie Homes. From *Where the Sky Began: Land of the Tall Grass* by John Madson. Copyright © 1982 by John Madson. Reprinted by permission of Houghton Mifflin Company.
3. Yellowstone Backcountry. Will and Jane Curtis.
4. Mount Assiniboine. Will and Jane Curtis.
5. The Columbian Ground Squirrel. Jane P. Curtis.
6. Caribou. Jane P. Curtis.
7. Adirondack Guide Boat. Jane P. Curtis.
8. The Cardiff Giant. Jane and Will Curtis.
9. Wine from the Finger Lakes. Jane P. Curtis.
10. St. Brendan. Jane P. Curtis.
11. Forgotten Viking Boat. Will and Jane Curtis.
12. Beatrix Potter. Jane P. Curtis.
13. Laplanders. Jane P. Curtis.
14. The Lipizzan Horses. Jane P. Curtis.
15. The Tower of Pisa. Will and Jane Curtis.
16. Sardinia. Jane P. Curtis.
17. Cordoba. Jane P. Curtis.
18. The Galapagos Islands. Jane P. Curtis.
19. Peruvian Textiles. Jane P. Curtis.
20. Inca Empire. Jane P. Curtis.
21. Quechua Indians. Will and Jane Curtis.
22. Llamas. Jane P. Curtis.
23. Peruvian Food. Jane P. Curtis.
24. Peru. Jane P. Curtis.
25. Mountain Trails. Will and Jane Curtis.
26. Dead Woman Pass. Jane P. Curtis.
27. Trek to Machu Picchu. Jane P. Curtis.
28. Cuzco. Jane P. Curtis.

PLANETS & SPACE

1. A Short History of All History. John H. Mitchell. *Sanctuary*, November 1982, Massachusetts Audubon Society.
2. The Winter Sky. Larry Prussin. Vermont Institute of Natural Science.
3. Arabian Astronomy. Farouk El-Baz. *Planetary Report*, March 1982.
4. Lightning. S. W. Foss. *New York State College Extension Service Bulletin*.
5. Space Trash. Linda Garvey. December 1981. Reprinted by permission of *SCIENCE 84 Magazine* © the American Association for the Advancement of Sciences.
6. Meteors and Meteorites. Larry Prussin. Vermont Institute of Natural Science.
7. Asteroids. Louis Friedman and Carl Sagan. *Planetary Report*, July/August 1983.
8. Getting Ready for Halley. Stephen P. Marteen. With permission from *Natural History*, Vol. 90, No. 12; Copyright the American Museum of Natural History, 1981.
9. Marsquake. "Giant Landslides in the Mariner Valleys on Mars." Conway Snyder, *Planetary Report*, June/July 1981.
10. Survival in Space. Bruce Webbon. With permission from *Natural History*, Vol. 90, No. 12; Copyright the American Museum of Natural History, 1981.
11. Aurora Borealis. Michael Caduto. Vermont Institute of Natural Science.
12. The Technicolor® Sky. Walter N. Webb. Courtesy of *Museum of Science Newsletter* (Boston), a Members publication. Summer 1982.
13. Project Sentinel. Thomas R. McDonough. *Planetary Report*, February 1984.

ENERGY

1. Energy and Agriculture, Extension Service, University of Vermont.
2. Solar Greenhouse. Sammi Izzo, Vermont Institute of Natural Science.
3. Wood Energy. Susanne Cockrell, Woodstock, Vt.
4. Drying Wood. Extension Service, University of Vermont.
5. Wind Power. Vermont Institute of Natural Science.
6. Wind-powered Generating Systems. "Wind Powered Generating Systems." Ibid.
7. Kerosene. Tim Matson. *Alternative Light Styles*, Countryman Press, 1984.
8. Efficient Lighting and Small Appliances. Extension Service, University of Vermont.
9. Coal. Elizabeth Kaufman and Bill Gilbert. Reprinted from *AUDUBON*, the magazine of the National Audubon Society; Copyright © 1978.

10. Oil-platform Fires. "Fighting Oil-Platform Fires At Sea." Horace S. Mazet. Reprinted by permission from *Sea Frontiers* © 1983 by the International Oceanographic Foundation, 3979 Rickenbacker Causeway, Virginia Key, Miami, Florida, 33129.
11. Arctic Mining. Melvin A. Conant. *Oceanus*, Winter 1982–3, Woods Hole Oceanographic Institute.

THE NUCLEAR SWORD

1. Reduction of the Ozone Layer. "Darkness at Noon: The Environmental Effects of Nuclear War." John Birks. *Sierra Magazine*, May/June 1983.
2. A Darkened World. Ibid.
3. A Full-scale Nuclear War? "What About the Children?" Parents and Teachers for Social Responsibility, Moretown, Vt.
4. What About the Children? Ibid.
5. Twisted Words, Distorted Images. Ibid.
6. The Ultimate Folly. Russell Peterson. November 1983. Reprinted from *AUDUBON*, the magazine of the National Audubon Society; Copyright © 1982.

Index

Adirondack Park, 210–211
Agriculture, and energy, 257–258
Alaska, Chilkat Valley, 126–127
American woodcock, 103–104
Antlers, 72–73
Ants, 164
Aphids, 163–164
Appliances, energy efficient, 266–267
Argali, 75
Armadillos, 92–93
Asteroids, 247–248
Astronomy: Arabian, 242–243; asteroids, 247–248; aurora borealis, 252–253; cosmology, 239–240; Halley's comet, 248–249; Mars, 249–250; meteors and meteorites, 246–247; space pollution, 245; sunlight, 253–254; winter sky, 240–241.
Aurora borealis, 252–253
Autumn: changing leaves in, 17–18, 52–53; evergreen leaf drop in, 50–52; flowers in, 46–47; garden in, 47–49; insulating houses in, 21–22; lawn care in, 53–54; ponds in, 19–20

Bag Balm, 68–69
Bats, brown, in winter, 26
Bears, grizzly, 204–206
Beavers, 69–70; in winter, 26
Bees: bumblebees, 152; honeybees, 63–64, 151–152; stings, 154–155; undertaking, 153
Beetles, 148–150; whirligig, 145
Big Dipper, 15
Bird houses, man-made, 111–112

Birds: adaptations of, 106–107; American woodcock, 103–104; bald eagles, 126–127; bitterns, 133–134; bluebirds, 117–118; common loons, 131–132; condors, 127–128; cuckoos, 119–120; feathers, 109–110; gannets, 140–141; great horned owls, 124–125; grouse, 122–124; gulls, 135–137; houses, 111–112; killdeer, 138–139; mobbing by, 113–114; molting, 110–111; nestlings, 114–115; nuthatches, 115–116; ostriches, 125–126; Passeriformes, 105; pine siskins, 121–122; protecting baby, 10; puffins, 137–138; migration, 107–108; swifts, 121–122; turkey vultures, 129–130; whip-poor-wills, 116–117; whooping and sand cranes, 134–135; winter feeding, 22–23, 115–116
Bitterns, 133–134
Black widow spiders, 166–167
Bluebirds, 117–118
Boats: Adirondack guide, 210–211; Viking, 215–216
Bogs, 179–180
Brendan, Saint, 214–215
Buffalos, 76
Butterflies, 159–162

Canyon de Chelly, 201–202
Cardiff giant, 211–212
Caribou, 209–210; migration of, 66
Cats, domestic, as predators, 78–79
Cheetahs, 90–91
Children, and nuclear war, 277–278

Chilkat Valley, Alaska, 126–127
Chipmunks, in winter, 26
Coal, 267–268
Compost, making, 43–44
Condors, 127–128
Coral, poisonous, 181–182
Cordoba, Spain, 222–223
Cougars, 89–90
Coy-dogs, 80
Coyotes, 79–80
Crabs, horseshoe, 190–192
Cranes, whooping and sand, 134–135
Crickets, 157–158
Cuckoos, 119–120
Cuzco, 236–237

Damselflies, 146
Dandelions, 11–12
Dead Woman Pass, Peru, 234–235
Deer: antlers, 72–73; in spring, 3–4; red, 73–74
Dew, 12–13
Dew point, 13
Dogs, as predators, 78–79
Dolphins, 94–95

Eagles, bald, 126–127
Egg Rock Island, Maine, 137–138
Eggs, amphibian, 6–7
1816, 1–2
Elephant, African, migration of, 65–66
Energy: agriculture and, 257–258; and arctic mining, 270–272; coal, 267–268; kerosene, 264; small appliances, 266–267; solar greenhouses, 258–259; wind, 262–264; wood, 259–260
Estuaries, 182–183
Evergreens, leaf drop of, 50–52

Fat, as biological insulation, 64
Fawns, protecting, 10
Feathers, 109–110
Finger Lakes region, wines from, 213–214
Fishers, 85–86
Flies: caddis, 146; crane, 145–146
Flowers: dandelions, 11–12; roadside, 34–35
Food, Peruvian, 230–231
Frogs, eggs, 7

Galapagos Islands, 224–225
Gannets, 140–141
Garden: in autumn, 47–49; Garibaldi's, 41–42; high-yield vegetables for, 42–43
Glaciers, and soil formation, 33–34
Gorp, 14
Grasshoppers, 156–157
Great Bear, 15
Greenhouses, solar, 258–259
Grouse, 122–124; in winter, 26
Gulls, 135–137
Guttation, 13

Habitat, definition of, 31
Halley's comet, 248–249
Hibernation, 26, 30
Hiking, 13–14
Himalayas, animals in, 74–75
Hippopotami, 95
Horses, Lipizzan, 219–220
Houses, early prairie, 202–204
Hull, George, 212

Ice age, and soil formation, 33–34
Incas, 226–237 *passim*
Insects: aphids, 163–164; bees, 151–155; beetles, 148–150; butterflies, 159–162; caddis flies, 146; and cold weather, 147–148; crane flies, 145–146; crickets, 157–158; damselflies, 146; grasshoppers, 156–157; lightning bugs, 155–156; mites, 51; pill bugs, 162–163; snowfleas, 28–29; in a stream, 144–146; structure of, 143–144; water striders, 145; whirligig beetles, 145. *See also* Spiders
Ireland, 214–216

Jaguarundi, 87–88

Kerosene, 265
Killdeer, 138–139
Kingfishers, 130–131
Kudzu, 39–40

Language, animals and, 63–64
Laplanders, 218–219
Lawn, autumn care for, 53–55
Leaves: changing colors of, 17–18,

52–53; as insulation, 21
Lighting, energy efficient, 266–267
Lightning, 243–244
Lightning bugs, 155–156
Lilacs, 8–9
Llamas, 229–230
Lobsters, blue, 188–189
Log houses, 202–203
Loons, common, 131–132

Machu Picchu, 231–236 *passim*
Maine, Egg Rock Island, 137–138
Mammals: antlers, 72–73; argali, 75; armadillos, 92–93; beavers, 69–70; biological insulation of, 64–65; buffalos, 76; cats, domestic, 78–79; cheetahs, 90–91; cougars, 89–90; coy-dogs, 80; coyotes, 79–80; dogs, 78–79; dolphins, 94–95; fishers, 85–86; Himalayan, 74–75; hippopotami, 95; hunter and hunted, 67–68; jaguarundis, 87–88; language and, 63–64; longevity of, 60–61; and man, 59–60; migration of, 65–66; mink, 83–84; mules, 77–78; penguins, 95; porcupines, 85–86; red squirrels, 71; river otters, 95; sea otters, 93–94; serows, 75; swimming adaptations, 94–95; tahrs, 75; water shrews, 86–87; weasels, 84–85; whales, 96–97; in winter, 26–30; wolves, 80–82; yaks, 75; yawning and, 62–63.
Manure, 45–46
Maple casebearers, 18–19
Maples: insects on, 18–19; red, 4–5
Mars, 249–250
Metamorphosis, insect, 144
Meteorites, 246–247
Meteors, 246–247
Mice, in winter, 26
Migration, 29, 65–66; spring bird, 107–108
Mildew, powdery, and lawns, 55
Milkweed, as monarch butterfly feed, 161
Mining, arctic, 270–271
Mink, 83–84
Mites, and evergreen needle drop, 51
Mobbing, 113–114

Moles, 3
Molting, 110–111
Monarch butterflies, 161–162
Moss and berry bowls, 56–57
Mount Assiniboine, 206–207
Mules, 77–78

Neurobiology, 173–174
New Mexico, 201–202
New Zealand, and red deer, 73–74
Northern lights, 252–253
Nuclear war, 276–277, 280–281; and children, 277–278; government views on, 279–281; and sunlight, 274–275; and ozone layer, 273–274
Nuthatches, 115–116

Oceanography, 193–194
Oil-platforms, fires on, 268–269
Ostriches, 125–126
Otters: river, 95; sea, 93–94
Owls, great horned, 124–125
Ozone layer, and nuclear war, 273–274

Palytoxin, 182
Panther, 90
Passeriformes, 105
Peat, 179–180
Penguins, 95
Peru, 231–232; Cuzco, 236–237; food in, 230–231; Incas, 226–237 *passim*; llamas, 229–230; Machu Picchu, 235–236; mountain trails, 233–235; Quechuas in, 228–229; textiles, 226–227
Pill bugs, 162–163
Pine siskins, 121–122
Pisa, Leaning Tower of, 220–221
Plants: poisonous, 37–38; terrariums, 56–57; watering indoor, 55–56
Pollution: from gas tanks, 175–176; in the home, 169–170; in space, 245
Ponds, in autumn, 19–20
Poplars, in spring, 4
Porcupines, 85–86
Potatoes, 230–231
Potter, Beatrix, 216–217
Precocialism, 114
Puffins, 137–138
Puma, 90

Index   295

Quechua Indians, 228–229

Raccoons: protecting baby, 10; in winter, 26

Salamanders, 183–185; eggs, 7
Salmon, 195–197
Sand dunes, 192–193
Sardinia, 221–222
Scythe, 16–17
Sea urchins, 187–188
Seahorses, 197–198
Seasons, 1–2
Sentinel Pass, 207
Serows, 75
Serviceberry, 8
Shadblow, 8
Shadbush, 7–8
Shells, 185–186
Shrews, water, 86–87
Shrimp, 198–199
Snow: benefits of, 23–24; as an insulator, 26–27
Snowfleas, 28–29
Snowmold, 54
Sod houses, 203–204
Soil, formation of, 33–34
Space, human survival in, 250–251; pollution in, 245
Spain, Cordoba, 222–223
Spiders, 165–167; black widow, 166–167; crab, 166
Spring, 3–4; salamanders in, 5–6; sounds of, 6–7; shadbush in, 7–8; trees in, 4–5
Springtails, 28–29
Squid, and neurobiology, 173–174
Squirrels: Columbian ground, 208–209; ground, 63; red, 71; in winter, 26
Starfish, 186–187
Stars: in summer, 15; in winter, 240–241
Stings, bee, 154–155
Storage of vegetables, 49–50
Summer: dew in, 12–13; stargazing in, 15
Sun, creation of, 240
Sunlight, 253–254
Swamps, bubbles in, 177–178
Swifts, 120–121

Tahrs, 75
Tamarack, in spring, 4–5
Terrariums, 56–57
Textiles, Peruvian, 226–227
Tidepools, 181–182
Toad eggs, 7
Trees: changing leaf colors, 17–18; evergreen leaf drop, 50–52; in spring, 4–5

Urubamba, 233, 235

Vegetables: harvesting, 48–49; storage of, 49–50
Velvet, antler, 72, 74
Viceroy butterflies, 162
Voles, 3
Vultures, turkey, 129–130

Wampum, 186
Water: pollution, 169–170, 175–176; woodlots and, 172–173
Water striders, 145
Water tables, 170–172
Weasels, 84–85
Weeds, 36–37
Wetlands, 176–177
Whales: beluga, 99; bowhead, 97–98; Greenland right, 97–98; hunting, arguments against, 96–97; white, 99–100
Whip-poor-wills, 116–117
Wildlife, protecting young, 9–10
Wind power, 262–264
Wines, New York State, 213–214
Winter: animals in, 26–30; benefits of snow in, 23–24; feeding birds in, 22–23; skygazing in, 240–241; snowshoes in, 24–25
Witch hazel, 46–47
Wolves, 80–82
Wood: drying, 261–262; energy, 259–260
Woodchucks, in winter, 26
Woodlots, and water, 172–173
Wrinkles, 189–190

Yaks, 75
Yawns, animals and, 62–63
Yellowstone Park, 204–206